Responding to Changes in Sea Level

Engineering Implications

Committee on Engineering Implications
of Changes in Relative Mean Sea Level

Marine Board
Commission on Engineering and Technical Systems
National Research Council

National Academy Press
Washington, D.C. 1987

NATIONAL ACADEMY PRESS • 2101 Constitution Avenue, NW • Washington, DC 20418

NOTICE: The project that is the subject of this report was approved by the Governing Board of the National Research Council, whose members are drawn from the councils of the National Academy of Sciences, the National Academy of Engineering, and the Institute of Medicine. The members of the panel responsible for the report were chosen for their special competences and with regard for appropriate balance.

This report has been reviewed by a group other than the authors according to procedures approved by a Report Review Committee consisting of members of the National Academy of Sciences, the National Academy of Engineering, and the Institute of Medicine.

The National Academy of Sciences is a private, nonprofit, self-perpetuating society of distinguished scholars engaged in scientific and engineering research, dedicated to the furtherance of science and technology and to their use for the general welfare. Upon the authority of the charter granted to it by the Congress in 1863, the Academy has a mandate that requires it to advise the federal government on scientific and technical matters. Dr. Frank Press is president of the National Academy of Sciences.

The National Academy of Engineering was established in 1964, under the charter of the National Academy of Sciences, as a parallel organization of outstanding engineers. It is autonomous in its administration and in the selection of its members, sharing with the National Academy of Sciences the responsibility for advising the federal government. The National Academy of Engineering also sponsors engineering programs aimed at meeting national needs, encourages education and research, and recognizes the superior achievements of engineers. Dr. Robert M. White is president of the National Academy of Engineering.

The Institute of Medicine was established in 1970 by the National Academy of Sciences to secure the services of eminent members of appropriate professions in the examination of policy matters pertaining to the health of the public. The Institute acts under the responsibility given to the National Academy of Sciences by its congressional charter to be an adviser to the federal government and, upon its own initiative, to identify issues of medical care, research, and education. Dr. Samuel O. Thier is president of the Institute of Medicine.

The National Research Council was organized by the National Academy of Sciences in 1916 to associate the broad community of science and technology with the Academy's purposes of furthering knowledge and advising the federal government. Functioning in accordance with general policies determined by the Academy, the Council has become the principal operating agency of both the National Academy of Sciences and the National Academy of Engineering in providing services to the government, the public, and the scientific and engineering communities. The Council is administered jointly by both Academies and the Institute of Medicine. Dr. Frank Press and Dr. Robert M. White are chairman and vice-chairman, respectively, of the National Research Council.

This report represents work supported by Cooperative Agreement Nos. 14-12-001-30301 and 14-12-0001-30228 between the Minerals Management Service of the U.S. Department of the Interior and the National Academy of Sciences.

Library of Congress Cataloging-in-Publication Data

National Research Council (U.S.). Committee on
Engineering Implications of Changes in Relative
Mean Sea Level.
Responding to changes in sea level.

Bibliography: p.
Includes index.
1. Coastal engineering—United States. 2. Sea
level—United States. I. National Research Council
(U.S.). Marine Board. II. Title.
TC223.N37 1987 627 87-21965
ISBN 0-309-03781-6

Printed in the United States of America

COMMITTEE ON ENGINEERING IMPLICATIONS OF CHANGES IN RELATIVE MEAN SEA LEVEL

ROBERT G. DEAN, *Chairman*, University of Florida
ROBERT A. DALRYMPLE, University of Delaware
RHODES W. FAIRBRIDGE, Columbia University
STEPHEN P. LEATHERMAN, University of Maryland
DAG NUMMEDAL, Louisiana State University
MORROUGH P. O'BRIEN, University of California, Berkeley
ORRIN H. PILKEY, Duke University
WILTON STURGES III, Florida State University
ROBERT L. WIEGEL, University of California, Berkeley

Government Liaisons

JACK E. FANCHER, National Oceanic and Atmospheric Administration
SUZETTE KIMBEL, Department of the Army
PAUL G. TELEKI, U.S. Geological Survey
JAMES G. TITUS, Environmental Protection Agency

Staff

MARTIN J. FINERTY, JR., Staff Officer, July 1985–January 1987
AURORA GALLAGHER, Senior Staff Officer to July 1985
ANDREA CORELL, Editorial Consultant
JOYCE B. SOMERVILLE, Administrative Secretary

THE MARINE BOARD

Preface

A large and growing proportion of the nation's population, facilities, and development is located along the Atlantic, Gulf, and Pacific coasts. Maintaining a sensitive balance of land with local sea levels may become more and more difficult in view of an increased rate of rise in worldwide sea level that has been predicted through the turn of the next century due to the projected global warming.

The National Research Council's Board on Atmospheric Sciences and Climate (BASC) and others have estimated that worldwide sea level may rise 70 cm (±25 percent) by the year 2075 as a result of thermal expansion, the melting of alpine and Greenland glaciers, and possible disintegration of the West Antarctic Ice Sheet (National Research Council, 1983). Such a rise would have significant implications for coastal communities and coastal engineering, the discipline that must address the problems of erosion and structural protection.

This study was initiated because of the potential consequences of such a rise to human life, communities, facilities, and the environment, and considering the large investment in developing coastal areas. The study's goal was to establish a basis for coastal planners, engineers, and government agencies to carry out their responsibilities in the presence of an anticipated increase in the rate of sea level rise over the next several decades. To focus this effort the Committee on Engineering Implications of Changes in Relative Mean Sea Level outlined the following objectives:

- examine knowledge concerning mean sea level changes;
- establish the rate of relative sea level change around the continental United States, based on tide gauge results;
- project sea level rise for engineering use over the next 100 years;
- examine the likely responses of sandy shorelines and wetlands to sea level rise;
- examine the potential effects on engineering works and built facilities;
- review measures for both preventing erosion and adapting to it; and
- determine whether new technologies are required.

Aspects of sea level change that do not have specific engineering implications, such as biological effects and land use planning, are outside the scope of the study, and are not directly discussed.

A committee of nine members was formed with expertise in coastal geomorphology, coastal engineering, and physical oceanography (see Appendix A). Representatives of four government agencies with specific interests in the subject were designated as liaison members to the committee. The principle guiding the constitution of the committee and its work, consistent with the policy of the National Research Council, was not to exclude members with potential biases that might accompany expertise vital to the study, but to seek balance and fair treatment.

The committee reviewed a variety of background materials from a variety of sources, including the International Geological Correlation Program 200 Project; Permanent Service for Mean Sea Level, Bidston Observatory, United Kingdom; Commission on Shorelines of the International Union of Quaternary Research; U.S. Environmental Protection Agency; and the National Research Council. The committee met several times over a 2-year period. Two workshops were also held in Washington, D.C. to review information presented by representatives of the private and public sectors, in addition to many other panel meetings over a 3-year-period. The issues addressed included climate and oceanic models, instrumentation for measuring sea level change, analysis of available records, potential contributors to change, economic implications of various scenarios of sea level rise, coastal drainage with rising sea level, coast protection techniques and structures, and physical processes of land loss in marshlands. A commissioned study was sponsored to investigate the effect on mean water levels in bays.

Based on committee activities and the professional experience of the members, the committee's conclusions and recommendations are presented in the Executive Summary and Chapter 9.

Contents

Responding to Changes in Sea Level

Engineering Implications

Executive Summary

Recently, several studies have projected increases in the rate of eustatic sea level rise, which some climatologists and oceanographers believe may accelerate due to a future warming of the atmosphere associated with the "greenhouse effect" produced by human-induced loading of the atmosphere with carbon dioxide and other gases. This interdisciplinary study of the engineering implications of relative mean sea level change examines

- recent sea level trends;
- projections of continuing relative change (over the next 100 years);
- shoreline response;
- consequences for engineering works and built facilities;
- methods for protecting structures from erosion and flooding, as well as adapting to shoreline retreat; and
- the need for new technologies for mitigation.

To provide a useful basis for sensitivity design calculations and policy decisions that must take sea level rise into account, the committee adopted three plausible variations in eustatic sea level rise to the year 2100, all displaying a greater rate of rise in the distant future than in the next decade and all with an increased rate of rise relative to the present: 50, 100, and 150 cm. To ensure efficient planning, it is important that these projections be kept up to date (see "Recommendations" herein).

Over the past century, worldwide sea level has risen about 12 cm. In many places along the U.S. shoreline, subsidence exceeds the eustatic component by a factor of 2, and in Louisiana by a factor of 10. In higher latitudes, glacial rebound is much greater than the eustatic component of sea level rise and, in locations like Hudson Bay, Canada, has resulted in a relative lowering of sea level in excess of 130 cm/century. These substantial differences must be considered in developing responses to a relative "change."

Early in this study the committee questioned whether sea level change trends based on tide gauges located inside bays and estuaries are representative of open-coast trends. A special study was commissioned by the Marine Board to address this question (Mehta and Philip, 1986).

The study concluded that (1) gauges located either inside or outside bays are subject to different influences that tend to degrade the quality of the data; (2) with more distant locations inside the bay, gauges contain a greater quantity of "noise" that is not representative of the outside mean sea level; and (3) in the short term (over several decades), inside gauges will underestimate rise taking place on the open coast. Certainly, the number of long-term, open-coast tide gauges should be increased with a special emphasis on the Southern Hemisphere, where present coverage is poor.

Over the next 25 years, the highest rate of sea level rise recommended for consideration in this report would produce a 10-cm rise. Although this may seem a relatively small increase, there are three situations in which the effects are greatly magnified:

1. Sandy beaches on the outer coast exposed to ocean waves where natural processes may cause beaches to erode 1 m or more for a 1-cm rise in sea level (Bruun rule).

2. The wedge of saline water through estuaries and tidal rivers may advance as much as 1 km for a 10-cm rise in mean sea level. This will be of special concern for drinking water supplies and coastal ecosystems during droughts. (Louisiana loses 100 km^2 [1 percent] of its wetlands for a 1-cm rise.)

3. Salinity intrusion in coastal aquifers where the landward displacement of the salt- and freshwater interface is a large multiplier of the sea level rise. Current problems of salinity intrusion

into groundwater supplies will be increased with only relatively small rises in sea level.

A significant increase in sea level could cause widespread shoreline erosion and inundation. The two general response options available are to:

1. stabilize the shoreline, either through beach nourishment or by new or augmented coastal armoring; or
2. retreat from the shoreline, maintaining a more-or-less equal elevation above local sea level.

Whether to defend or to retreat depends on several factors including the future sea level rise rate and the cost of retreat. The former is poorly known and the latter will vary from place to place.

Potential problems associated with sea level change can be categorized into two classes: those of the open coast where both water level and wave action are concerns, and those of inland tidal waters where wave action is usually much less severe. Wave action effects are so complete and potentially so devastating that they require special expertise for structural design. As a result of erosion along the open coast, structures not designed for such forces may become exposed to wave action.

Design procedures for coastal structures should include a review of data on past water levels, including the maximum level, and should then provide some margin of safety to cover uncertainties. In some cases (e.g., docks), structures conservatively designed with expected lives of 50 years or less should not be significantly affected by sea level rise even if a rise is not considered specifically in the design. Other structures, such as sea walls and hotels on the open coast, would be vulnerable to even a small rise.

Where possible, considerations of sea level changes should be incorporated into coastal land-use planning. Areas designated primarily for industrial use may not be significantly affected by required coastal protection. However, designated uses contingent on the continuation of existing environmental features, such as shoreline conditions, may limit shoreline response options. For example, coastal armoring of a formerly sandy beach may reduce the environmental desirability of the area.

Construction of almost any conceivable protection against sea level rise can be carried out in a relatively short period of time. Therefore, if a substantial increase should occur, there will be time to implement protective measures. However, in areas where such protection would not be justified, a cost-effective abandonment of facilities would require decades to implement. Throughout the world, works exist that are applicable for protection against the effects of a rise in relative mean sea level. These options should be reviewed by engineers, planners, and policymakers.

The committee concluded that the most appropriate present engineering strategy is not to adopt one particular sea level rise scenario, but instead to be aware of the probability of increasing sea level and to keep all response options open. In many engineering projects, it may be desirable to carry out sensitivity calculations, using specific sea level rise scenarios. If a particular structure is ill-suited for retrofitting, it will undoubtedly be appropriate to allow for an acceleration of sea level rise in the initial design. Sea level change during a structure's design service life should be considered along with other factors, but a change does not present such essentially new problems as to require new techniques of analysis.

The committee's recommendations highlight the need for continuing and increased scientific study of the rates and causes of sea level change and the development of a sound basis for forecasting these changes. Efforts to understand coastal processes and the effects of sea level rise on engineering projects should also be expanded considerably.

A concise listing of the committee's specific conclusions and recommendations follows.

CONCLUSIONS AND RECOMMENDATIONS

Conclusions

1. Relative mean sea level, on statistical average, is rising at the majority of tide gauge stations situated on continental coasts around the world. Relative mean sea level is generally falling near geological plate boundaries and in formerly glaciated areas such as Alaska, Canada, Scandinavia, and Scotland. Relative mean sea level is not rising in limited areas of the continental United States, including portions of the Pacific Coast.

2. The contrasting signals concerning relative mean sea level behavior in different parts of the United States (and the world in general) are interpreted as due to differing rates of vertical motion

of the land surfaces. Subsidence and glacial rebound are significant contributors to vertical land displacements.

3. Large, short-term (2–7 year) fluctuations worldwide are related to meteorological phenomena, notably shifts in the mean jet-stream path and the El Niño-Southern Oscillation mechanisms, which lead to atmospheric pressure anomalies and temperature changes that may cause rise or fall of mean sea level by 15–30 cm over a few years.

4. Studies of a very small number of tide gauge records dating more than 100 years (the oldest being Amsterdam, started in 1682) show that after removal of the subsidence factor where known, mean sea level has been fluctuating through a range of not more than 40–150 cm (in long-term fluctuations) for at least 300 years.

5. The geological record over the last 6,000 years or so indicates that there has been a general, long-term rise with short-term fluctuations probably not exceeding 200 cm during the last 1,500 years.

6. Monitoring of relative mean sea level behavior is at present inadequate for measuring the possible global result of future climate warming due to rising greenhouse gases. The most serious gaps in present tide gauge coverage are in three areas: (a) high polar latitudes, (b) midoceanic locations, and (c) the entire Southern Hemisphere.

7. Because of localization of many extreme subsidence processes, especially those connected with anthropogenic extraction of fluids such as groundwater and hydrocarbons, tide gauges are needed at every major coastal city to gather data to assist in evaluating the long-term regional trend of relative mean sea level.

8. The risk of accelerated mean sea level rise is sufficiently established to warrant consideration in the planning and design of coastal facilities. Although there is substantial local variability and statistical uncertainty, average relative sea level over the past century appears to have risen about 30 cm relative to the East Coast of the United States and 11 cm along the West Coast, excluding Alaska, where glacial rebound has resulted in a lowering of relative sea level. Rates of relative sea level rise along the Gulf Coast are highly variable, ranging from a high of more than 100 cm/century in parts of the Mississippi delta plain to a low of less than 20 cm/century along Florida's west coast.

9. Accelerated sea level rise would clearly contribute toward

a tendency for exacerbated beach erosion. However, in some areas, anthropogenic effects, particularly in the form of poor sand management practices at channel entrances, constructed or modified for navigational purposes, have resulted in augmented erosion rates that are clearly much greater than would naturally occur. Thus, for some years into the future, sea level rise may play a secondary role in these areas.

10. As noted previously, the two response options to sea level rise are stabilization and retreat. Retreat is most appropriate in areas with a low degree of development. Given that a "proper" choice exists for each location, selecting an incorrect response alternative could be unduly expensive.

11. There does not now appear to be reason for emergency action regarding engineering structures to mitigate the effects of anticipated increases in future eustatic sea level rise. Sea level change during the design service life should be considered along with other factors, but it does not present such essentially new problems as to require new techniques of analysis. The effects of sea level rise can be accommodated during maintenance periods or upon redesign and replacement of most existing structures and facilities. There are very limited geographic areas where current subsidence rates may require near-term action as has been the case in Japan and Terminal Island, California.

12. When not restrained by funding, availability of materials, or work force, construction of almost any conceivable protection against sea level rise can be carried out in a very short time; short, that is, relative to the rate of sea level rise.

13. Defensive or mitigative strategies are site specific and cannot be developed nationwide on the basis of a blanket generalization or comprehensive legislation.

Recommendations

1. The prognosis for sea level rise should not be a cause for alarm or complacency. Present decisions should not be based on a particular sea level rise scenario. Rather, those charged with planning or design responsibilities in the coastal zone should be aware of and sensitized to the probabilities of and quantitative uncertainties related to future sea level rise. Options should be kept open to enable the most appropriate response to future changes

in the rate of sea level rise. Long-term planning and policy development should explicitly consider the high probability of future increased rates of sea level rise.

2. The three previously described scenarios of sea level rise used in this study (see Figure 2-2) provide a useful range of possible future sea level changes for design calculations. The general shape of these curves is concave upward with greater rates of rise in the distant future than those in the next decade or so. The confidence that these scenarios will encompass the actual levels decreases with increasing time, and significant deviations outside the range of these scenarios are possible, including an amelioration in the rate of rise. Thus, the committee recommends that these projections be updated approximately every decade to incorporate additional data and to provide an improved basis for planning and response to the rise.

3. Practitioners can more readily incorporate the implications of sea level rise if probabilities reflecting uncertainties are attached to the projections. Thus, it is recommended that appropriate statistical techniques be applied to develop a probability distribution associated with sea level rise through the year 2100 and that all updated projections include such information.

4. Feasibility studies for coastal projects (e.g., shore protection projects of the U.S. Army Corps of Engineers and storm surge studies of the Federal Emergency Management Agency) should consider the high probability of accelerated sea level rise. It may be some time before precise estimates of future sea level rise are possible. In the meantime, the risks associated with a substantial rise should not be disregarded. Instead, feasibility studies should consider which designs are most appropriate for a range of possible future rates of rise. Strategies that would be appropriate for the entire range of uncertainty should receive preference over those that would be optimal for a particular rate of rise but unsuccessful for other possible outcomes.

5. The federal government should acquire long-term reliable accurate data from a water-level measuring system for open-ocean stations at scientifically important locations throughout the world. Critical stations should include documentation of vertical ground motion and the temporal salinity and temperatures of the water column. Tide gauges should be installed at every major coastal city.

6. The important decision for maintaining or abandoning

coastal facilities in the face of rising sea level should be well documented by scientific knowledge. Agencies that fund coastal research, such as the U.S. Navy, U.S. Army, National Science Foundation, National Oceanic and Atmospheric Administration, U.S. Geological Survey, and the Environmental Protection Agency, should increase their funding for coastal processes research. The federal research funding effort should focus on studies directed toward understanding nature's response to relative sea level rise and developing appropriate engineering responses. A substantial portion of this research should be conducted at universities and other laboratories and centers throughout the coastal United States to ensure the development of requisite engineering capability in regions of the country where it will be most helpful.

1
Relative Mean Sea Level

A significant portion of the world's population lives within the coastal zone, with many buildings and facilities built at elevations less than 3 m (10 ft) above mean high-tide level along the shoreline. Without any secular change in the height of mean sea level or in the height of the land, the elevations of structures are not adequate to ensure the safety of people and works in the event of major storms (including storm surges), especially when such events coincide with infrequent but predictable perigean spring tides.

This hazard has grown increasingly apparent and serious along many of the world's coastlines as local mean sea level has risen during the twentieth century. Although in some areas of the world the local sea level is falling, the predominant change is a rising sea level with rates ranging from 1 to 5 mm/yr. International attention has been drawn toward this problem by two possibly interrelated sets of observations: (1) relative mean sea level is rising and beach erosion is being exacerbated in many parts of the world (Bird, 1985), including many areas of the United States (Figure 1-1); and (2) the atmospheric level of "greenhouse gases" is steadily rising as a result of the combustion of fossil fuel and deforestation. While it is tempting to correlate the two, a cause-and-effect (eustatic) response has not been proven to date, although a future eustatic response is a clear possibility (Barth and Titus, 1984).

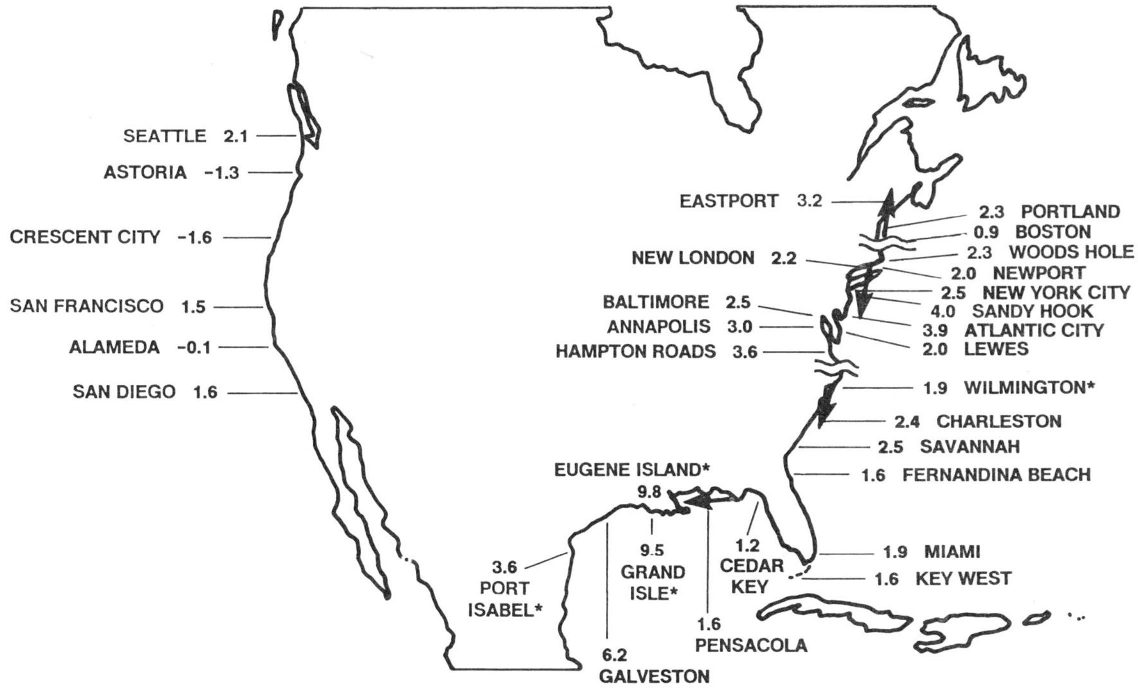

FIGURE 1-1 A summary of the present best estimates of local relative sea level changes along the U.S. continental coastline in mm/yr. The figures are based on the tide gauge records over different intervals of time during the period 1940–1980. Much regional variability is evident. Source: Adapted from Stevenson et al. (1986).

Carbon dioxide and the other trace gases that comprise "greenhouse gases" create a greenhouse effect in the troposphere. The combined effects are still poorly understood but seem likely to establish a sequence of climatic effects that could result in a general global warming (National Research Council [NRC] 1983, 1982, 1979). The result will be an increased rate of rise in glacioeustatic and steric-expansion eustatic sea level.

Relative mean sea level change at a particular location is the difference between the eustatic (global) change and any local change in land elevation. The long-term causes of relative mean sea level rise are sixfold, but not all of the processes are operative in every locality.

1. *Eustatic rise* of world sea level. "Eustatic" means a global change of the oceanic water level. Its most important forms at the present time are regarded as glacio-eustasy, caused by melting of land-based glacier ice, and the steric expansion of near-surface ocean water due to global ocean warming. "Steric" refers to the specific volume of the medium, which expands when heated or shrinks when cooled.

2. *Crustal subsidence* or uplift of the land surface due to neotectonics, that is, contemporary, secular, structural downwarping of the earth's crust. Tectonic phenomena occur in five distinctive categories: subsidence of former glacio-isostatic marginal uplift belts (e.g., the eastern United States); cooling crustal belts following rifting (e.g., parts of the Gulf of California); subsidence in regions of long continued sediment loading (e.g., East and Gulf coasts, especially the Mississippi delta); uplift in regions of active crustal subduction (e.g., Puget Sound); and subsidence due to loading by volcanic eruptions (e.g., Hawaii, Aleutians Islands).

3. *Seismic subsidence* of the land surface due to sudden and irregular incidence of earthquakes.

4. *Auto-subsidence* due to compaction or consolidation of soft, underlying sediments, especially mud or peat.

5. *Man-made subsidence* due to structural loading, as well as groundwater, and oil and gas extraction. Of the four subsidence processes only this category, anthropogenic subsidence, can be reversed or at least partially mitigated by recharge or other management actions.

6. *Variations due to climatic fluctuations* are a consequence of oceanic factors including El Niño-Southern Oscillation (ENSO)

effects, and are related to secular changes in the size and mean latitudes of subtropical high pressure cells. Along mainland coasts (especially east coasts in the Northern Hemisphere), a decreasing current flow associated with warming epochs causes a rise in sea level due to the Coriolis effect, whereas in midoceanic gyre regions there is no mean sea level change. This issue requires study (Cartwright et al., 1985; Barnett, 1983b). It also appears that tide gauge records contain substantial long-period fluctuations (5–100 years), which indicate that the accurate extrapolation of small sea level rise values from the data is very difficult. Furthermore, determining changes in rates of rise is even more difficult.

Of these identified causes of sea level rise, only the eustatic rise is a universal, global effect (by definition). For any one area the other causes come into play in various proportions. It should be stressed that no national survey of the local extent of the processes has ever been undertaken, but it is clear that the variations will be highly regional.

Various segments of the U.S. coastline experience subsidence or uplift due to factors (2) through (5). Superimposed on this regional subsidence is the global eustatic sea level rise. If the greenhouse effect/glacier melt concept is confirmed, its potential contribution to mean sea level rise will outstrip other causes of relative sea level rise along most of the U.S. coastline by 2025.

PAST CHANGES IN RELATIVE MEAN SEA LEVEL

Geologic Record of Sea Level

It has been established that during the last Ice Age (15,000 years ago), mean sea level was perhaps as much as 100–150 m lower than it is now. Sea level rose rapidly until about 6,000 years ago, when the rate of change of global sea level became quite low compared to that earlier period of time. An assessment developed by Shepard (1963) is presented in Figure 1-2. During the past 6,000 years there were perhaps fluctuations one or more meters over a thousand years (or more) apparent in some areas (Fairbridge, 1961).

In regions of very rapid subsidence (e.g., Mississippi delta, Rhine delta), eustatic trends tend to be obscured (Van de Plassche, 1986). In contrast, the formerly glaciated regions of the world

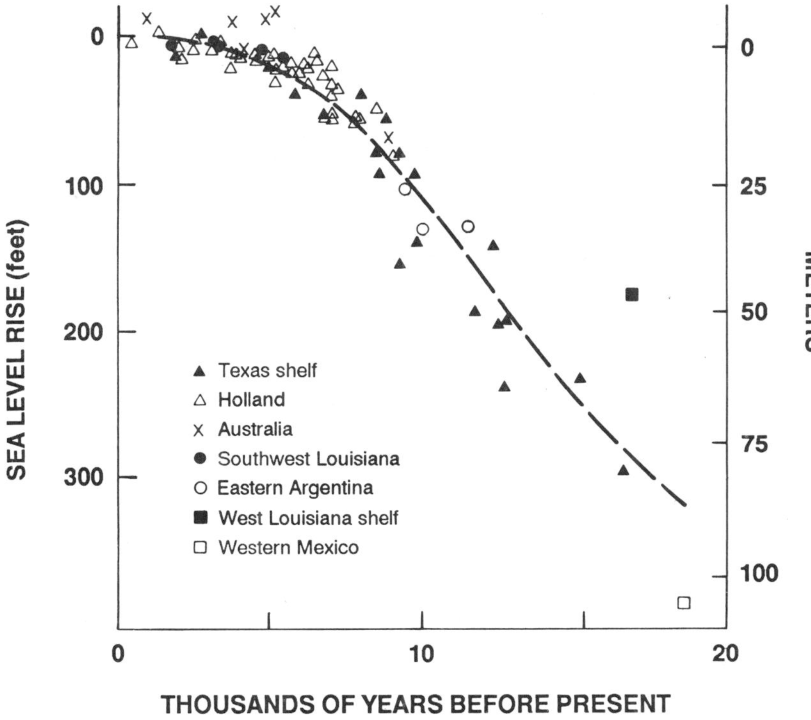

FIGURE 1-2 Sea level elevations versus time as obtained from carbon 14 dates in relatively stable areas. Source: Adapted from Shepard (1963).

(notably, most of Canada, Scandinavia, and Scotland) are regions of neotectonic tilting or uplift, where rates of uplift have exceeded past rates of sea level rise.

Sea Level Rise in the Twentieth Century

For nearly a century, relative mean sea level has maintained a steady rise at many tide-recording stations around the world. At the same time the atmospheric greenhouse gases have shown a steady rise, and, more recently, other greenhouse gases such as chlorofluorocarbons have been increasing. To some observers the increase of greenhouse gases in the atmosphere implies a warming of the world's climate, although the evidence to date is still being debated by the scientific community (NRC, 1983, 1982) and most experts do not expect the warming to be detected until the 1990s.

A tide gauge simply records the current sea level at a particular location. Since sea level is a basic consideration for short-range coastal construction plans, designers of long-lived coastal structures should consider the change in sea level that may occur during the structures' useful life. For this purpose, simple projection of the trend of local tide gauge records is inadequate; the underlying causes of sea level change must be addressed and techniques employed to forecast the effect of each cause.

Tide gauge data are available for approximately the past century. However, these data provide the sea level relative to the supporting base of the gauge, and that base may be either sinking or rising. For example, Figure 1-3 presents results from long-term tide gauge records at Atlantic City, New Jersey; San Francisco, California; and Juneau, Alaska. The approximate recorded relative rates of change are

- Atlantic City: + 0.40 m/century,
- San Francisco: + 0.13 m/century, and
- Juneau: – 1.38 m/century,

where a positive rate indicates a relative rise. The relative drop in sea level at Juneau reflects the rebound (rise) of the land with unloading of the land following the melting of glacial ice.

A question raised early in the development of this study was whether sea level change trends based on tide gauges located inside bays and estuaries are representative of open-coast trends. This concern was posed as early as 1929 in a report sponsored by the National Research Council (Johnson, 1929). Specifically, the issue is whether the effects of engineering works, primarily channel deepening and the construction of jetties for navigational purposes, would affect the sea level rise trend measured by bay gauges.

A special study commissioned by the Marine Board to address this concern includes a summary of analytical relationships and empirical results relevant to this problem (Mehta and Philip, 1986). Mehta and Philip concluded that gauges located inside and outside bays are each subject to different influences that tend to degrade the quality of the data. With locations more distant inside the mouth of the bay, gauges contain a greater amount of "noise," which is not representative of the open-coast sea level. Analytical

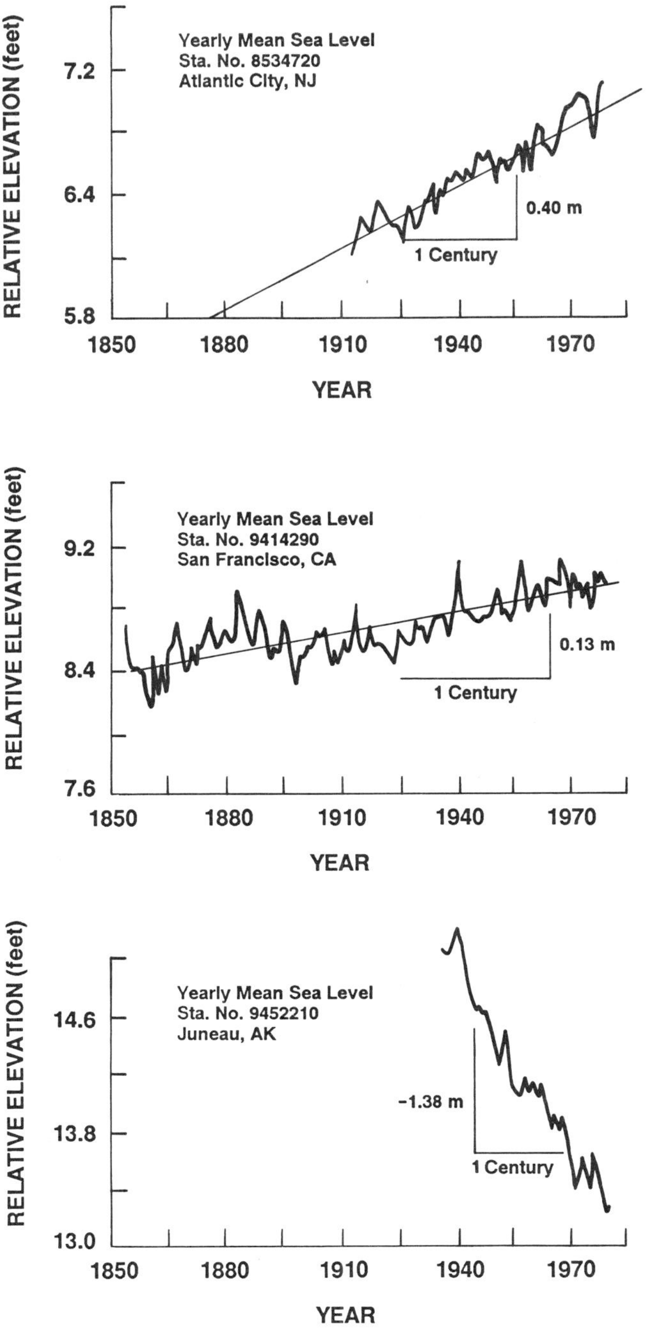

FIGURE 1-3 Tide gauge data for (a) Atlantic City, New Jersey; (b) San Francisco, California; and (c) Juneau, Alaska. Source: Hicks et al. (1983).

TABLE 1-1 Difference in the Secular Change of Mean Sea Level for Selected Gauge Pairs (outside minus inside)

Gauge Pair Outside	 Inside	Secular Change Difference (mm/yr) Outside - Inside
Long Branch, N.J.	New York, N.Y.	13.1
Atlantic City, N.J.	New York, N.Y.	1.4
Duck Pier, N.C.	Norfolk, Va.	2.1
Springmaid Pier, S.C.	Charleston, S.C.	13.6
Daytona Beach Shores, Fla.	Mayport, Fla.	6.4
Key Colony Beach, Fla.	Fort Myers, Fla.	2.4
Vaca Key, Fla.	Fort Myers, Fla.	2.2
Key West, Fla.	Miami, Fla.	-0.3
Naples, Fla.	Fort Myers, Fla.	0.7
Clearwater Beach, Fla.	St. Petersburg, Fla.	-0.6
Cedar Key, Fla.	Pensacola, Fla.	-0.4
Shell Point, Fla.	Pensacola, Fla.	1.8
La Jolla, Calif.	San Diego, Calif.	-0.1
Santa Monica, Calif.	Los Angeles, Calif.	0.8
Ricon Island, Calif.	Los Angeles, Calif.	3.2
Monterey, Calif.	San Francisco, Calif.	-3.9
Arena Cove, Calif.	San Francisco, Calif.	-13.1
Trinidad, Calif.	San Francisco, Calif.	4.8
Crescent City, Calif.	Astoria, Ore.	-0.3

NOTE: A positive change denotes a higher outside than inside rate.

SOURCE: From Hicks as reported by Mehta and Renji (1986).

considerations suggest that any short-term (several decades) differences will be biased toward a lower trend from this inside gauge compared to outside gauges.

In addition, Hicks' (1984) results of long-term sea level trends were assessed from pairs of gauges inside bays versus open-coast gauges. The comparison included 19 gauge pairs, with each outside gauge selected as that in closest proximity to the inside gauge (Table 1-1). The average trend difference of the 19 gauge pairs was 1.8 mm/yr with the outside rate exceeding that inside. When the three absolute differences exceeding 10 mm/yr are excluded, the average difference decreases to 1.1 mm/yr. Thus, relative to outside gauges, this study indicates that the trend rate from inside gauges has underestimated somewhat the relative mean sea level rise rate.

TABLE 1-2 Estimates of Eustatic Sea Level Rise per Century Based on Tide Gauge Data

Author	Estimated Rise (cm)
Thorarinsson (1940)	> 5
Gutenberg (1941)	11 ± 8
Kuenen (1950)	12 to 14
Lisitzin (1958)	11.2 ± 3.6
Fairbridge and Krebs (1962)	12
Hicks (1978)	15 (United States only)
Emery (1980)	30
Gornitz et al. (1982)	12[a]
Barnett (1983a)	15

[a] Ten centimeters excluding long-term trend.

SOURCE: Adapted from Barnett (1983a) and Hicks (1978).

Range of Sea Level Estimates Available

Using tide gauge results from around the world, various estimates have been obtained for mean rate of change (Table 1-2). The large variance is partly the result of gross geographical imbalance of the gauge sites; most are in the Northern Hemisphere midlatitudes, almost all are on continental shores, and almost none are in high latitudes. A map of gauge locations is presented in Pugh and Faull (1983).

The lack of insular tide gauges, until recently, has deprived analysts of any means of testing mean sea level change for the Coriolis effect. Gauges located toward the middle of an oceanic gyre that is speeded or slowed will fall or rise respectively, whereas those on continental shores show an opposing trend. The lack of a sufficient number of high-latitude gauges precludes testing for planetary spin-rate effects (although there is astronomical proof of changes in the earth's spin rate which should be registered by sea level, especially at high latitudes). It is noted that the spin rate has decreased slightly and would result in relative mean sea level decreases and increases at low and high latitudes, respectively. This has been proposed, but not confirmed, as an indicator of antarctic melting. It is not clear whether this signal could be isolated in the presence of the other components and "noise."

Furthermore, many gauges are located in obviously subsiding delta areas, and others in tectonically rising areas.

The first attempt to filter out the grossly anomalous data of world tide gauge records was done by Fairbridge and Krebs (1962); their results indicated a mean rise for the first part of the twentieth century of 1.2 mm/yr. Subsequent analyses, using different filtering procedures but progressively larger data sources, showed similar results; for example, Lisitzin (1974), 1.12 mm/yr, and Gornitz et al. (1982), 1.2 mm/yr. Several analysts have suggested a change during the last few decades; Barnett (1983a) showed for 1930–1980 an average of 2.3 mm/yr, and Emery (1980) gave a value of 3.0 mm/yr.

The newest global results are by Pirazzoli (1986, 1984). Of 1,178 records provided by the Permanent Service for Mean Sea Level, 229 stations were selected as having > 50-year records (plus a few with 30 years) of consistent trends (Figure 1-4). Only 13 percent showed the "mean eustatic" value (1.0–1.5 mm/yr); 22.5 percent showed 1.0–2.0 mm/yr; 20.5 percent showed a rise of 0.1–1.0 mm/yr. Pirazzoli indicates that the extreme variance between the stations emphasizes the importance of local subsidence. Furthermore, he stresses that there is no unequivocal demonstration of any eustatic rise at all, at least during the last 40–50 years. Gornitz (unpublished data) has prepared averages for a number of mid-Pacific atolls: over 20–30 years, Nawiliwili shows a mean rise of 0.30 mm/yr; Canton Island, 0.31 mm/yr; Eniwetok, 0.81 mm/yr; and Midway, minus 1.34 mm/yr. For regional studies (e.g., Australia), Aubrey and Emery (1983) were unable to identify "unambiguously" a eustatic signal.

Munk et al. (1985) indicate that phenomena that create large amplitude fluctuations in sea level, such as the ENSO, make it very difficult to obtain statistically reliable estimates of rates of change in mean sea level and even more difficult to detect whether the rate of change is increasing or decreasing.

In summary, the gauge measurements, in a few cases continuing over 100–300 years after correction for known trends, suggest mean sea level fluctuations that are generally consistent with the geological record of the past 6,000 years (Ters, 1986; Tooley, 1978; Fairbridge, 1961). The nature of the related climatic changes, insofar as it has been possible to document them, is also consistent.

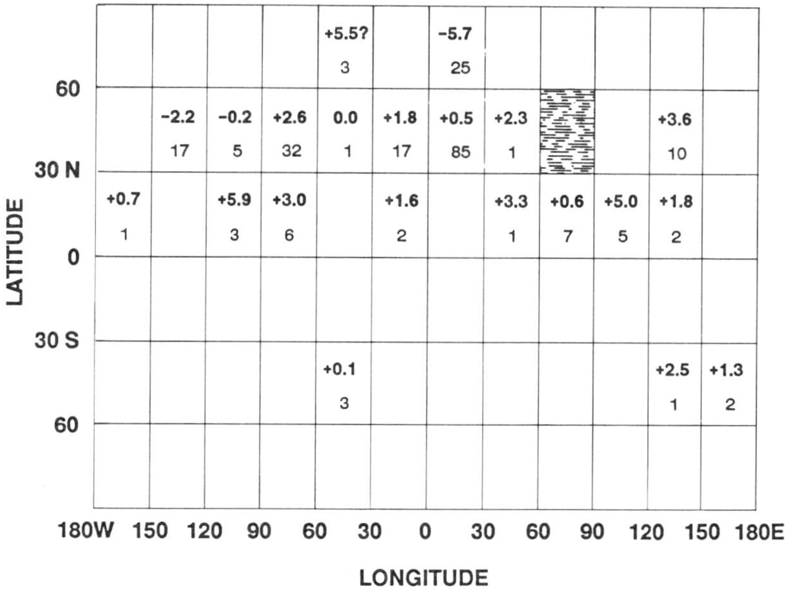

FIGURE 1-4 Geographical distribution of worldwide change of relative mean sea level (by latitude/longitude coordinates) using tide gauge records extending more than 50 years (augmented with a few of 30 years). Large figures and mathematic signs indicate average relative rise or fall in mm/yr; small figures the number of stations employed. Note that 97.4 percent of the data is from the Northern Hemisphere. The shaded box is the only rectangle (Central Asia) with no marine coastline. No data exist for 70 percent of the boxes, and 70 percent of the stations are located in only four boxes. It is evident that no statistically valid basis exists for assuming that eustatic rise is in progress, at least from the data presented here. Nevertheless, it is important that five boxes with very low rates of change are midoceanic stations. Four boxes with values over 5 mm/yr are located in areas of recognized crustal uplift or subsidence. To identify and quantify a global eustatic rate requires not only more and better-selected gauging points, but also more-sophisticated geological and oceanographic analysis. Source: Pirazzoli (1984).

Additional examination is required. The record of climatic fluctuation indicates quite appreciable variations (Lamb, 1984; Wright, 1983).

METHODS OF OBSERVING RELATIVE MEAN SEA LEVEL

With reference to any one locality, relative mean sea level is measured by different techniques that fall within the expertise of different groups of specialists, who for the most part do not meet professionally. The techniques are as follows:

1. *Tide gauge analysis* is processed globally by a commission of the International Association of Physical Oceanography and nationally in the United States by the National Ocean Service (NOS). Monthly relative sea level maps of the tropical Pacific are issued by the University of Hawaii.
2. *Satellite altimetry* is administered by the National Aeronautics and Space Administration (NASA) in the United States and studied internationally by groups within the International Union of Geology and Geophysics (IUGG). Satellite imaging can also be employed for geomorphic analysis.
3. *Geodetic leveling* is globally reported by the Commission for Recent Coastal Movements of the International Association of Geodesy (IAG, which adheres to IUGG) and in the United States by the National Geodetic Survey.
4. *Geomorphological and geological analysis* is globally coordinated by the Commission on Shorelines and the Commission on Neotectonics of the International Union for Quaternary Research (INQUA). There is a U.S. National Committee for INQUA with individual representation on the commissions but no official governmental participation.

There is also the Sea Level Project of the International Geological Correlation Programme (No. 61, completed in 1984; followed by No. 200, led by P. Pirazzoli, CNRS-Intergeo Geographical Institute, Paris), which also has a U.S. committee with individual membership.

Each of these organizations collect data in time series that refer to specific locations. However, the time series cover very different intervals. The relevant geological studies deal with about the past 6,000 years; the others concentrate on periods ranging from two or three centuries to a month or less. Nevertheless, from all these data, regional and global means are obtained by varied and sophisticated statistical analysis. While good agreement is possible for the regional means of relative mean sea level (coastal sectors of 500–1,000 km), there is so far no unanimity as to the global values for eustatic sea level rise. Efforts are now being directed at defining the eustatic components, and to delineating and understanding the local and regional effects.

Of the various observational techniques, tide gauge analysis furnishes the most detailed, accurate, and directly measured record of relative mean sea level, but its hour-by-hour variability is subject to complex disturbances (e.g., by atmospheric pressure or local rainfall), and the data therefore require extremely careful analysis. The results provide valuable monthly, seasonal, annual, and more-than-decadal trends.

Tide gauge data provide basic input for the design of coastal structures. For short-lived structures, the annual tide ranges and the mean tide level are sufficient; however, for longer-lived coastal structures, the long-term trends become important in the design process. Satellite altimetry is potentially very valuable, but requires a longer base period (Wyrtki, 1985). Recent availability of the Global Positioning System (GPS) provides absolute (not relative) sea level to centimeter accuracy. Future contributions of this new technology, supplemented by traditional measurement techniques, should be extremely beneficial to understanding eustatic sea level trends and local neotectonics.

Geodetic leveling involves the periodic releveling of the first-order vertical topographic survey stations, about once every two or three decades, and thus provides evidence of secular deformation of the earth's crust, as well as indications of local compactional subsidence. The method is extremely time consuming, and although very successful in Japan, Scandinavia, and Eastern Europe, it has been poorly funded in the United States and many of the available data have not been analyzed. The analyses provide decadal to century-long trends. The Japanese have developed and applied ingenious methods of sensing local and short-term compaction rates.

Finally, geomorphological and geological analysis furnishes century to millennial trends. This method is particularly valuable in demonstrating the natural, long-term response of any area or region, a trend or pattern of behavior that can then be compared with the short-term time series provided by geodetic leveling and tide gauges, which may contain extraneous or anomalous short-term data. The short-term data should be evaluated in terms of known long-term trends whenever possible.

Two Examples

How these varied approaches can be constructively integrated

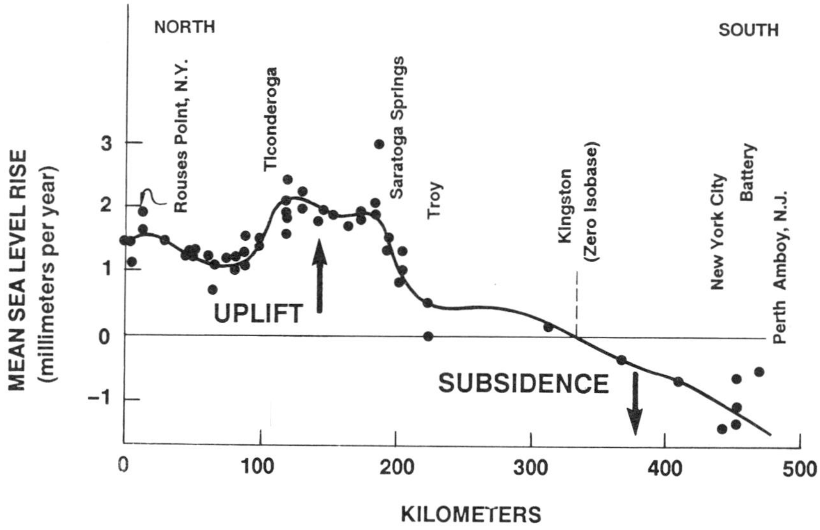

FIGURE 1-5 Geodetic leveling profile from Perth Amboy, New Jersey through New York City to Rouses Point, New York on the Canadian border. Mean sea level rise at the Battery, New York City, averaged 2.7 mm/yr. Subtracting 1.2 mm/yr as the eustatic component leaves 1.5 mm/yr representing subsidence and long-term oceanographic factors. Source: U.S. Coast and Geodetic Survey data, 1902 to 1955; adapted from Fairbridge and Newman (1968).

is illustrated by reviewing data from two well-documented sites: New York City and Long Beach, California.

The tide gauge of New York City is located at the Battery and set on hard crystalline bedrock that is not disturbed by sediment compaction, frost action, or human action, such as groundwater withdrawal. For 90 years the record shows a systematic (though fluctuating) rise of mean sea level of about 2.7 mm/yr (Hicks et al., 1983). Deducting about 1.2 mm/yr as the eustatic component, 1.5 mm/yr remains as a probable crustal subsidence factor. This has been checked by geodetic and geologic techniques.

The geodetic leveling lines were followed up the Hudson River to the Canadian border and filtered to remove highly deviant data points (Figure 1-5). A secular tilting of the crust during the present century is shown, with the Canadian border area rising at about 1–2 mm/yr and the New York City area sinking at about the same rate; a null point is situated near Kingston (Fairbridge and

Newman, 1968). The geologic surveys show that over the last 6,000 years the southern end of the section has been sinking slowly and the Canadian end has been rising. About 9,000 years ago, Lake Champlain lay at sea level. This is a clear-cut case, demonstrated by independent data sets, that the New York City area is subsiding at about 1–2 mm/yr. The lowering of land affects the bedrock of the whole region and is not affected by human activity.

In California, despite frequent earthquakes, the tide gauge records are remarkably stable and coherent from station to station. San Francisco has the longest series (since 1855), which shows that if interannual variations are removed, mean sea level shows broad fluctuation but has generally risen at 1.3 mm/yr over 125 years (Hicks et al., 1983). No distinguishable change is evident for the last several decades.

In contrast to San Francisco and New York, Long Beach Harbor, California commenced a sudden, substantial, relative sea level rise trend in the 1950s, submerging appreciable parts of Terminal Island, which is in the harbor. This subsidence is anomalous when compared with long-term trends and was diagnosed as a short-term local phenomenon that is related to the withdrawal of oil, natural gas, and water during exploitation of the Wilmington Oilfield. Artificial recharge of the porous strata has slowed the rate of continued subsidence, but the cost of dike building and other land preservation measures exceeded $100 million.

The New York City and Long Beach examples demonstrate that areas must be considered in three contexts: local, regional, and global.

2
Assessment of Changes in Relative Mean Sea Level

ESTIMATES OF FUTURE MEAN SEA LEVEL RISE

During the past 4 years, several groups have attempted to estimate the rise in eustatic mean sea level that could result from projected global warming. Studies have focused on four factors: the thermal expansion of ocean water; the melting of mountain glaciers; the melting of Greenland glaciers; and the possibility that antarctic glaciers may slide into the oceans. To estimate the significance of these processes also requires an estimate of future global warming, which in turn depends on future concentrations of "greenhouse gases" and the sensitivity of the climate to changes in these concentrations.

Long-term carbon dioxide monitoring stations are situated to minimize localized effects of industrialization. Records dating back to 1958 from the top of Mauna Loa, Hawaii are presented in Figure 2-1. They indicate that in addition to substantial seasonal variations, there is an upward trend in the concentration of greenhouse gases, with an increase from 315 to 340 ppm occurring from 1958 to 1981. A similar increase was detected at an antarctic observation station. In addition to these documented increases, an increase of 50 ppm since preindustrial times (around 1850) has been estimated using tree ring data.

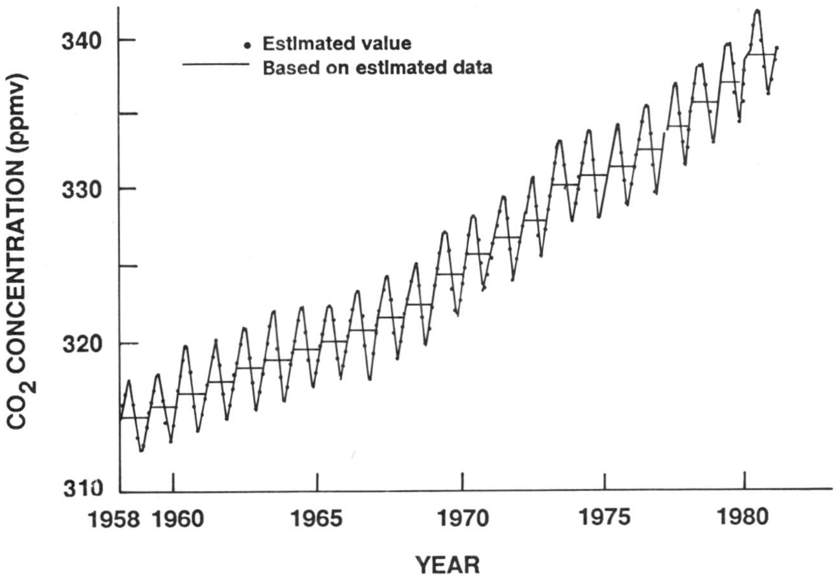

FIGURE 2-1 Mean monthly CO_2 concentration at Mauna Loa, Hawaii. Source: NRC (1983).

In *Carbon Dioxide and Climate: A Scientific Assessment* (NRC, 1979), it was concluded that a doubling of CO_2 would raise the earth's average surface temperature 1.5–4.5°C, with the warming at the poles two to three times as great as the average warming. The panel concluded: "We have tried but have been unable to find any overlooked physical effect that could reduce the currently estimated global warming due to a doubling of CO_2 to negligible proportions."

Nordhaus and Yohe (NRC, 1983) estimated the likely rate of increase for CO_2 considering uncertainties regarding future energy use patterns, economic growth, and the ability of the oceans to absorb carbon dioxide. The panel estimated that there is a 98 percent probability that CO_2 concentrations will be at least 450 ppm (1.7 times the preindustrial level) by the year 2050, and a 55 percent chance that the concentration will be 550 ppm. They estimated the probability of doubling CO_2 concentrations by 2100 as 75 percent. Ramanathan et al. (1985) estimate that the combined impacts of methane, chlorofluorocarbons, nitrous oxide, and several other trace gases will be approximately comparable to the warming resulting from greenhouse gases buildup.

On the basis of a global warming of 3–4°C, Revelle (1983) assessed the likely rise in sea level assuming that no antarctic deglaciation takes place. He estimated that Greenland and mountain glaciers could each contribute 12 cm to sea level in the next century, and that thermal expansion could contribute 30 cm. Based on current trends, Revelle concluded that other factors could contribute an additional 16 cm, for a total rise of 70 cm, with an estimated uncertainty of ±25 percent.

Hoffman et al. (1983) developed a variety of sea level rise scenarios based on high and low assumptions for all the major uncertainties. Although they used a fairly sophisticated model for projecting global warming and thermal expansion, they cautioned that the absence of glacial process models kept them from making accurate projections of snow and ice contributions to sea level. They estimated that sea level was likely to rise between 26 and 39 cm by the year 2025 and between 91 and 137 cm by 2075.

The report *Glaciers, Ice Sheets, and Sea Level* (NRC, 1985b) provided the first detailed look at the possible glacial contribution to sea level rise. Meier (1984) estimated that a global warming of 1.5–4.5°C could lead to a 10–30 cm alpine contribution to sea level rise. Bindschadler (1985) estimated a similar contribution from Greenland. There was less consensus regarding the antarctic contribution. Thomas (1985) estimated that the antarctic contribution resulting from a 4°C warming would most likely be 28 cm, but could be as high as 2.2 m. The panel concluded that the antarctic contribution was likely to be a few tenths of a meter, and possibly as great as 1 m or as little as a 10 cm drop. The panel did not estimate the likely contribution of thermal expansion (see discussion below).

Hoffman et al. (1986) revised their earlier projections in light of the glacial process models provided by the Polar Research Board (NRC, 1985b) and new information on future concentrations provided by the Board on Atmospheric Sciences and Climate (NRC, 1983) and Ramanathan et al. (1985). Although the revised assumptions had a minor impact on the estimates of thermal expansion, they substantially lowered the estimates of snow and ice contributions until after 2050. Hoffman et al. estimated the rise by 2025 to be between 10 and 21 cm, and by 2075 to be between 36 and 191 cm (Tables 2-1 and 2-2).

TABLE 2-1 Contributions to Future Sea Level Rise in the Year 2100 (in centimeters)

Study	Thermal Expansion	Alpine Glaciers	Greenland	Antarctica	Total
Hoffman et al. (1986)	28-83	12-37	6-27	12-220	57-368
Thomas (1985)	--	--	--	0-220	--
Hoffman et al. (1983)	28-115	b	b	b	56-345
NRC (1983)	--	10-30	10-30	-10- +100	--
Revelle (1983)[a]	30	12	12	c	70

[a] Contributions in the year 2085.

[b] Hoffman et al. assumed that the glacial contribution would be one to two times the contribution of thermal expansion.

[c] Revelle attributes 16 cm to other factors.

TABLE 2-2 Temporal Estimates of Future Sea Level Rise (in centimeters)[a]

Study	Year 2000	2025	2050	2075	2085
Hoffman et al. (1986)					
Low	3.5	10	20	36	44
High	5.5	21	55	191	258
Hoffman et al. (1983)					
Low	4.8	13	23	38	--
Mid-range low	8.8	26	53	91	--
Mid-range high	13.2	39	79	137	--
High	17.1	55	177	212	
Revelle (1983)[a]	--	--	--	--	70

[a] Other studies only provided an estimate for a specific year.

Based on an examination of the various component processes, Robin (1986) has proposed and applied a linear correlation between rise in mean global sea level, ΔSL, and mean global temperature, ΔK, that is

$$\Delta SL(t) = F \Delta K(t - t_o),$$

where F is a linear correlation coefficent and the time t_o represents the lag of sea level in responding to temperature change. Robin notes that although nonlinear correlation is more appropriate, the data are not sufficiently well conditioned to quantify adequately the coefficients modifying the nonlinear terms. Additionally, over temperature changes of interest, the sea level response should be nearly linear. Based principally on an analysis of the Gornitz et al. (1982) results, the following range of correlation coefficients was developed

$$16 \text{ cm}/^\circ\text{C} < F < 30 \text{ cm}/^\circ\text{C}.$$

Applying these values to an estimated temperature change of 3.5 ± 2°C due to a doubling of greenhouse gases over the next century, the associated range in global mean sea level change is from 24 to 154 cm.

All of the studies have focused on changes in the world's average sea level. However, as discussed previously, there is reason to believe that the relative rise will be somewhat greater along most of the U.S. coast. In Atlantic City, for example, relative sea level has risen 40 cm in the last century (Hicks et al., 1983), while the global rise has been estimated at 10–15 cm (Barnett, 1983a; Gornitz et al., 1982). Assuming that the processes responsible for local and regional subsidence do not change, over the next century the rise in sea level at Atlantic City and some of the U.S. Atlantic and Gulf coasts will be 10–30 cm greater than the global average.

SCENARIOS USED IN THIS REPORT

Because the rate of future sea level rise is uncertain, there must be uncertainties in any assessment of the implications. For its analyses to reflect these uncertainties, the committee examined three possible scenarios of eustatic sea level rise to the year 2100: rises of 0.5 m, 1.0 m, and 1.5 m.

The total relative sea level change above present levels at time t, $T(t)$, is the sum of the local, L, and eustatic, E, components and is expressed as $T(t) = L(t) + E(t)$. One possible equation for the eustatic contribution, the form of which appears generally consistent with anticipated future sea levels and which will be adopted for this study, is

$$E(t) = 0.0012t + bt^2,$$

in which $E(t)$ is the additional eustatic component in meters above present levels and t is the time in years from present.

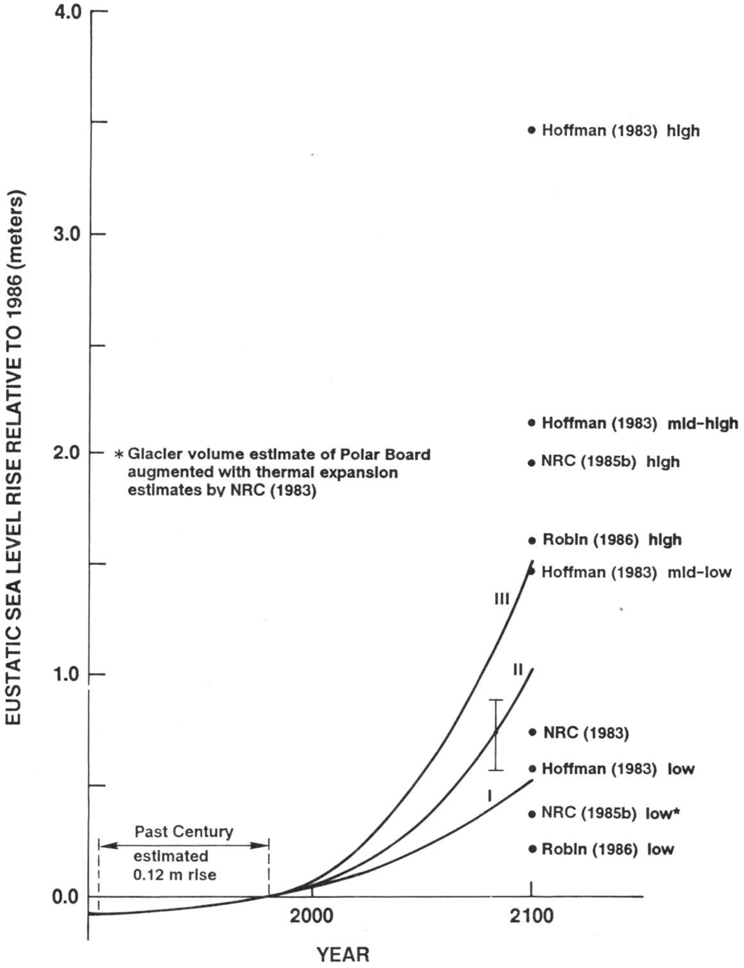

FIGURE 2-2 Eustatic sea level rise scenario adopted in this report compared with other estimates.

This component is presented in Figure 2-2 for the three scenarios considered, which results in eustatic components at the year 2100 of 0.5 m, 1.0 m, and 1.5 m, respectively. Also shown are estimates developed in various studies. The local component varies greatly from subsidence to uplift. The total component is $T(t) = (0.0012 + M/1,000)t + bt^2$ in which M (in mm/yr; $M = dL/dt$) is obtained from Table 2-3, and b for each of the three scenarios is presented in Table 2-4.

TABLE 2-3 Estimate of M, the Local Subsidence (+) or Uplift (-) Rates as Determined from Tide Gauge Measurements[a]

Location	Rate (mm/yr)	Location	Rate (mm/yr)
Portland, Maine	1.0	Grand Isle, La.	8.9
Boston, Mass.	1.0	Sabine Pass, Tex.	12.0
Providence, R.I.	0.5	Galveston, Tex.	5.1
Montauk, N.Y.	0.6	Port Isabel, Tex.	2.1
New York, N.Y.	1.5	St. Georges (Bermuda)	1.4
Atlantic City, N.J.	2.9		
Philadelphia, Pa.	1.4	San Diego, Calif.	-0.4
Lewes, Del.	1.9	Los Angeles, Calif.	-0.4
Baltimore, Md.	2.0	San Francisco, Calif.	0.1
Hampton Roads, Va.	3.1	Crescent City, Calif.	-1.7
Wilmington, N.C.	0.6	Astoria, Ore.	-1.3
Charleston, S.C.	2.2	Seattle, Wash.	0.8
Mayport, Fla.	1.0	Juneau, Alaska	-13.8
Miami Beach, Fla.	1.1	Skagway, Alaska	-19.5
Key West, Fla.	1.0	Honolulu, Hawaii	0.4
St. Petersburg, Fla.	0.8	Apra Harbor, Guam	-2.8
Pensacola, Fla.	1.1	Wake Island	-0.1

[a]The eustatic component of sea level change has been treated as steady (1.2 mm/yr) in developing this table. In reality, this effect is global but it may not be steady.

SOURCE: Based on Hicks (1983).

TABLE 2-4 Value of Coefficient b for Three Scenarios Considered in This Report

Scenario Eustatic Component by Year 2100 (m)	b (m/yr^2)
I. 0.5	0.000028
II. 1.0	0.000066
III. 1.5	0.000105

3
Relative Sea Level Rise Effects and Responses

As sea level rises, the shoreline may respond by flooding or erosion. The magnitude of these effects will depend on a number of factors, including whether the shoreline is on the open ocean or in a protected bay, the effect of any stabilizing structures, and population density and activities.

The related shoreline changes will have economic and environmental impacts that will require decisions to define a proper response. For example, for sandy shorelines, the possible remedies to erosion range from shoreline stabilization (through beach restoration or protective structures) to retreat from the shoreline. Each is technically feasible, but the appropriate response will usually be dictated by a combination of economic and environmental concerns.

In the case of retreat, planning and timing are important elements of the process. Given that a definite proper choice exists on a site-by-site basis, it follows that an inappropriate choice could be very expensive. For example, continued shoreline stabilization of an area of high erosional tendency and low economic base could be too costly. On the other hand, if the natural system is only slightly out of balance and a large tax base exists, a decision to retreat could be inappropriate and unduly costly. This latter condition would certainly hold for port cities, which have considerable investments in infrastructure.

Thus, there is a need for development and application of methodologies for estimating the expected erosion impact and attendant storm damage. With such a framework available and the conveyance to the public of plans for responding to a major problem (rebuilding or retreating), a much more rational response could be developed for implementation. This would clearly be of value to those responsible for allocating future financial resources and directing responses in times of high emotions, losses of personal property, and possible losses of life.

The decision to retreat or rebuild following a major storm would be a matter of the gravest concern. Many situations can be envisioned in which the temptation would be to make the wrong choice. For example, suppose that a storm of 500-year-return frequency were to strike an area, causing widespread damage and major change. Without the knowledge that this was indeed a very rare storm, the enormity of the damage and emotional impact could cause decision makers to be swayed toward retreat. On the other hand, if a relatively frequent storm (say, with a return period of 40 years) causes major damage due to the relatively high vulnerability of an area, retreat may be the appropriate response.

The development of a rational decision-making framework requires an understanding of both complex physical coastal processes and the economics of the area. Chapters 4–6 of this report review some of the necessary knowledge of coastal processes and the influence of sea level rise.

Historically, U.S. efforts to cope with relative sea level rise have been limited in scope and modest compared to those in some European countries. Until 1929, when the U.S. Army Chief of Engineers appointed the Board on Sand Movement and Beach Erosion to carry out field studies of shore processes, there was no organized effort to study the engineering problems of ocean shores. The board was appointed to study the processes at work, not to solve particular erosion problems. The U.S. Army Corps of Engineers had responsibility for improving and maintaining navigation works of all kinds, including the tidal entrances to coastal harbors. In 1930, the U.S. Congress authorized a successor and permanent organization, the Beach Erosion Board, now the Coastal Engineering Research Center.

From 1929 to the beginning of World War II, the Beach Erosion Board and a few universities worked on coastal processes and engineering problems. Support was limited and progress was slow. Beginning about 1940, and continuing after the war, amphibious operations, offshore petroleum production, coastal siting of electric power stations, the growth of coastal communities, environmental and ecological interest, and other developments caused a tremendous growth in the support for coastal studies and projects and university programs. The number of professional coastal experts has increased proportionately with the growth of these programs.

The engineering of works on the ocean coasts deals primarily with the characteristics and effects of wind-generated, ocean surface waves. Expertise in this area is the essential qualification of the coastal engineer. Because of this, categorization of engineering problems should be made on the basis of "waves" or "no waves," and this differentiation should extend to the effects of a sea level rise; those involving wave action differ fundamentally from those that do not.

Before discussing the effect of rising sea level on specific types of coastal projects, there are some principles that qualified coastal engineers generally accept as valid:

1. Structures are expensive and the ocean is a relentless adversary. A development "set-back" line may be the only action justified for undeveloped or lightly developed shores.
2. Sandy coastal shores are made of natural units, such as the length between inlets or a beach terminated by headlands, and must be treated as such. The effect of any structure anywhere in one of these units on the remainder of the shoreline must be analyzed before construction, and the plan should provide for mitigation of adverse effects, if any.
3. Sand tends to collect in sheltered areas around coastal structures such as groins or jetties. Plans should provide for predicting the location and capacity of such areas and for filling them initially with sand from other than the active shoreline, so as not to deplete other portions of the sand system.
4. The direction and magnitude of littoral transport is the most uncertain feature in the plans for a coastal project. Site-specific data are difficult and expensive to obtain, but it is extremely risky to proceed without them.
5. The choice of coastal structure for erosion mitigation will depend on site-specific factors. Structures that work satisfactorily in one location may prove to be totally inadequate (and perhaps detrimental) in another application.

4
Affected Hydrodynamic Processes

Normal and extreme events, including day-to-day wave action, storm surges, and high waves attending northeasters and hurricanes, affect the stability of the shoreline and can threaten the integrity of coastal works and upland structures. Obviously, a rising relative sea level, with resultant higher storm tides and larger waves, can only increase these hazards. The magnitude of some hydrodynamic processes will be affected significantly; others will be relatively unaffected. The mechanics governing these hydrodynamic processes are discussed briefly below, along with approximate estimates of magnitudes of change.

STORM SURGE

Storm surges, the flooding induced by wind stresses and the barometric pressure reduction associated with hurricanes, tropical storms, and northeasters, will be modified by sea level rise mostly in areas of very mild offshore slopes. With higher sea levels the larger expanse of shallow water will result in increased storm surge elevations compared to areas of steep offshore slopes, because the surge heights are proportional to both the length and the inverse slope of the offshore bottom. However, if the shoreline is fixed and the offshore water depths increase, then (referenced to the quiescient water level) the storm surges will be less, as the surge also varies inversely with the absolute water depth. This reduction in surge height for a uniform depth offshore profile is proportional to the ratio of the change in sea level to the initial water depth. As an example, if the return periods of storm surges resulting in water levels of 3 and 4 m are 50 and 100 years, then to a first approximation, a sea level rise of 1 m would result in an increase in frequency of the 4 m surge level from 100 years to 50 years. Expressed differently, the probability that a water level of 4 m would occur in a 50-year period would increase from 0.40 to 0.64, an increase of 62 percent.

A more complete consideration will demonstrate that for a shelf of uniform depth, the wind stress component of surge would decrease. To illustrate, consider an idealized continental shelf of uniform depth, η_o. As shown by Dean and Dalrymple (1984), the maximum storm surge, $\eta_{\max}$, at the shore is

$$\frac{\eta_{\max}}{h_o} = \sqrt{1 + 2A} - 1,$$

in which A is a ratio relating the magnitude of wind stress terms to the hydrostatic force terms. The factor A varies inversely with the water depth. It can be shown that the change in storm surge relative to the increased sea level is actually decreased by the amount

$$\frac{\Delta\eta_{\max}}{S} = \frac{-\eta_{\max}/h_o}{1 + \eta_{\max}/h_o}.$$

As an example, for a sea level increase of $S = 1$ m, a representative water depth of $h_o = 10$ m, and a wind stress surge of $\eta_{\max} = 3$ m, the reduction of storm surge relative to the increased sea level is

$$\Delta\eta_{\max} = -0.23 \text{ m}.$$

Of course, relative to an absolute datum, the storm surge (including the effect of sea level rise) is increased by 0.77 m. The above treatment considers only the wind stress component. The barometric depression component of storm surge is generally about 20–30 percent of the total and is weakly dependent on the water depth through relatively complicated dynamics that depend on the relative speeds of hurricane system translation and a free long wave.

The Federal Emergency Management Agency (FEMA) determines base flood elevations for the coastal counties of the United States. These elevations include the still-water level flood elevations, which have a 100-year return interval. Additionally, FEMA predicts the 100-year wave heights, which are superimposed on the base flood elevations. With sea level rise, the Flood Insurance Rate Maps (FIRMs) will need to be adjusted periodically to ensure that the premiums charged coastal property owners are actuarially sound. Since the design life of a house is likely to be 50 years, the FIRMs should perhaps be recomputed approximately every two decades. An alternative would be to allow 10 or 20 years of predicted relative sea level rise, including local subsidence, to be incorporated into the rate maps.

The predicted sea level rise will be manifested in two different ways—the change in surge elevation and the change in wave heights felt at the shoreline. The present methodology used by FEMA is to determine the wave heights at the shoreline based on a breaking condition; that is, the shoreline wave height is 78 percent of the water depth at the flooded shoreline. With rising sea level the offshore water depth will be greater, and as these storm waves propagate inland they will be larger than before.

TIDAL RANGES AND CURRENTS

In sheltered embayments, such as gulfs, bays, estuaries, and lagoons, the increase in sea level will be felt predominantly through an increase in water level. The depth increase will allow tidal waves to propagate faster due to the depth dependence. However, many of these coastal areas may have sedimentation rates commensurate with the relative sea level rise, resulting in minimum change in tidal characteristics.

For tidal bays and lagoons, the increase in mean sea level will result in increased tidal prisms (the volume of water carried into bays from low to high water by the tidal currents). This increase in prism will be due to an increase in the bay planform area (as a result of inundation and shoreline retreat) and by a reduction of friction in tidal entrances (because of deeper water). For sandy coastlines, O'Brien (1969) has shown that there is an equilibrium relationship between the tidal prism of a bay and the cross-sectional area of the entrance. Therefore, as the tidal prism increases, the tidal entrance will increase in area. Ramifications are that inlets controlled by jetties will become deeper, with implications about the stability of jetties and adequacy of bridge clearance. Uncontrolled inlets may deepen and widen, or new inlets may be created. However, the change in cross section is likely to be small, as estimated by the following argument. From O'Brien (1969), the inlet cross-sectional area A_c is related to the tidal prism P as

$$A_c = \mathrm{B}P^{\mathrm{m}},$$

where B and m are empirical constants. Taking derivatives and dividing,

$$\frac{\mathrm{d}A_c}{A_c} = \mathrm{m}\frac{\mathrm{d}P}{P}.$$

The percentage change in cross-sectional area is directly proportional to the percentage change in tidal prism. The constant m is on the order of unity. The change in prism due to inundation can be shown to be related to the perimeter (or shoreline length) C of the bay and the relative sea level rise S:

$$P = \frac{T_r C S}{2s},$$

where T_r is the tide range and s is an average shoreline slope. In general, the change in tidal prism is very small, when compared to the total tidal prism.

It is more difficult to make an assessment for the open coast. There is likely not to be significant change in tidal ranges and amplitudes; however, there is very little data on which to base a firm conclusion. If the tidal ranges at the shorelines are forced by the deep-ocean tides, then almost no changes in coastal tides will result. However, in some regions such as the Gulf of Maine, where the tidal dynamics are near resonance, relatively small changes in sea level could have significant effects on tidal heights and currents.

WAVES

It has been argued that with rising sea level the continental shelf will deepen, thereby resulting in less wave damping and higher wave energy at the shoreline. Additionally, due to this greater water depth and associated reduction in bottom friction, wave generation will be enhanced. These two problems will be addressed below as cases A and B.

Case A

This case applies to a preexisting wave of height $H(0)$, propagating across a continental shelf of width l having uniform water depth h. The equation governing wave damping across the shelf (in the x-direction) is

$$\frac{\delta(EC_g)}{\delta x} = -\overline{\tau_b U},$$

in which E is the wave energy per unit surface area, C_g is the wave group velocity, τ_b is the bottom stress, and U is the wave-induced water particle velocity just outside the bottom boundary layer. Using linear wave theory, the equation can be integrated to yield the wave height $H(x)$ at some location x, in terms of the initial wave height $H(0)$:

$$H(x) = \frac{H(0)}{1+F},$$

where

$$F \equiv \frac{2C_f \sigma^3 H(0)x}{3\pi g C_g sinh^3 kh},$$

and k is the wave number ($= 2\pi$/wave length), g is gravity, C_f is the bottom stress coefficient in the relationship ($\tau_b = \rho C_f/U/U$), ρ is the mass density of water, and σ is the wave angular frequency ($= 2\pi$/wave period). (See also Dean and Dalrymple, 1984.) From this equation, larger water depths decrease the size of F and hence decrease the amount of wave height reduction due to friction. This effect becomes more important for wider continental shelves.

The increased wave height $H'(l)$ due to sea level rise just outside the breaker zone can be expressed in terms of the corresponding wave height prior to sea level rise as

$$\frac{H'(l)}{H(l)} = \frac{1+F}{1+F'}.$$

As an example, consider the following as somewhat representative along the East Coast of the United States:

- Depth, $h = 10$ m
- Shelf width, $x = 10$ km
- Wave period, $T = 8$ s

- Initial wave height, $H(0) = 2$ m
- Friction coefficient, $C_f = 0.01$
- Sea level rise, $S = 1$ m.

For this case, the equations yield

$$\frac{H'(l)}{H(l)} = 2[\frac{1+0.143}{1+0.112}] = 2.054,$$

or about a 3 percent increase in wave height due to the water depth increase. This small increase is not likely to cause changes of substantial engineering significance.

Case B

Consider wave generation across a continental shelf. Wave growth will be enhanced by deeper water (due to sea level rise) because of the reduced effect of bottom friction. An estimate of this effect can be obtained through the shallow water forecasting relationships provided in the *Shore Protection Manual* (U.S. Army Corps of Engineers, 1984). For the case of a very long fetch (the distance over which the wind blows) and shallow water, the equation can be expressed as

$$\frac{gH}{w^2} = 0.15\left(\frac{gh}{w^2}\right)^{0.75},$$

where w is the wind speed. An increase in water depth S gives the following change in the wave height H:

$$\frac{\Delta H}{S} = 0.75\frac{H}{h}.$$

For the same values as in the last example:

$$\Delta H = 0.15 \text{ m},$$

or a 7.5 percent increase in wind-generated wave height as a result of the movement of the offshore region due to sea level rise. The effects of reduced wave damping and augmented wave generation would be combined in an approximate linear manner.

Larger wave heights in the surf zone will result in greater amounts of sediment movement, as most transport formulas include wave height to some power, and greater wave forces and potential for overtopping.

5
Effects of Sea Level Rise in the Coastal Zone

Sea level rise will have different effects along various portions of the U.S. coastline depending on conditions such as sediment type and coastal planform. It is possible to divide the coasts into physiographic regions for consideration of their response to relative sea level rise. For instance, the conditions in Louisiana do not apply to the coast of Maine because the Mississippi delta region is very flat, undergoing pronounced compaction and subsidence, while northern New England is characterized by nonerodible cliffs and portions are experiencing neotectonic uplift.

The present rise in water level is a complex phenomenon, including local, regional, and global components, as detailed previously. Shoreline position will respond to the cumulative effect of vertical motions, termed the relative mean sea level rise, regardless of their cause. However, it is instructive to divide the coasts into regions that will behave in a similar manner due to particular processes and materials.

The U.S. continental coastline is highly variable in character but certain regional trends are apparent. Tectonic mapping clearly indicates the reasons for the prominent differences between the Pacific coastal range as compared to the Atlantic and Gulf coastal plains. The tectonically active Pacific rim is a coast where one plate is being subducted below another (Inman and Nordstrom,

1971), resulting in a narrow continental shelf and an essentially nonexistent coastal plain. This coast is characterized by headlands and intervening pocket beaches. By comparison, the Atlantic and Gulf coasts have long been tectonically stable and constitute trailing edge and marginal sea coasts, respectively.

The Atlantic coastal plain, which extends over 100 miles inland, is characterized by a gently sloping surface with gradients of only several feet per mile near the shore. The prominent landforms from Long Island, New York to Miami Beach, Florida along the Atlantic coastal plain are barrier islands. The Gulf coastal plain exhibits the lowest average relief and gentlest gradients. Barrier islands are again the dominant coastal landforms, but the chain is less continuous than on the Atlantic Coast, as broken by the Mississippi River deltaic sediments of Louisiana, the marshy outcrops along the northeast Gulf coast of Florida (Tanner, 1960), and the broad outflowing of the Everglades along the limestone rocky coast of the southern peninsula of Florida.

These three coasts can be further subdivided into physiographic regions on the basis of geologic history and coastal morphodynamics. Basically 11 types of coasts can be defined for the U.S. continent (Figure 5-1) using a modification of the classification by Shepard and Wanless (1971). The glaciated coast extends from northern New Jersey to Maine. This physiographic region can be further subdivided into the erosion-resistant crystalline (granite) rock of northern New England and the mostly unconsolidated glacial till of southern New England. While coastal barriers have developed by spit growth across many embayments, the mainland is often fairly high near the shore, forming cliffs. The low areas, which are subject to storm surges, are clearly marked on topographic maps.

The coastal compartment barrier chains (Swift, 1968) of New Jersey and the Delmarva Peninsula are characterized by four sectors: (1) terminal north spit, (2) low, eroding headland, (3) long barrier islands backed by open lagoons, and finally to the south (4) short, stubby barriers with marsh-filled embayments. Within these compartments, there is variable shoreline vulnerability in response to differing rates and patterns of shoreline erosion (Leatherman et al., 1982), storm surge flooding, and inlet breaching potential. The south shore of Long Island, New York can also be included in this sector since the glacial outwash plain there has a similar physiography to that of the Atlantic coastal plain.

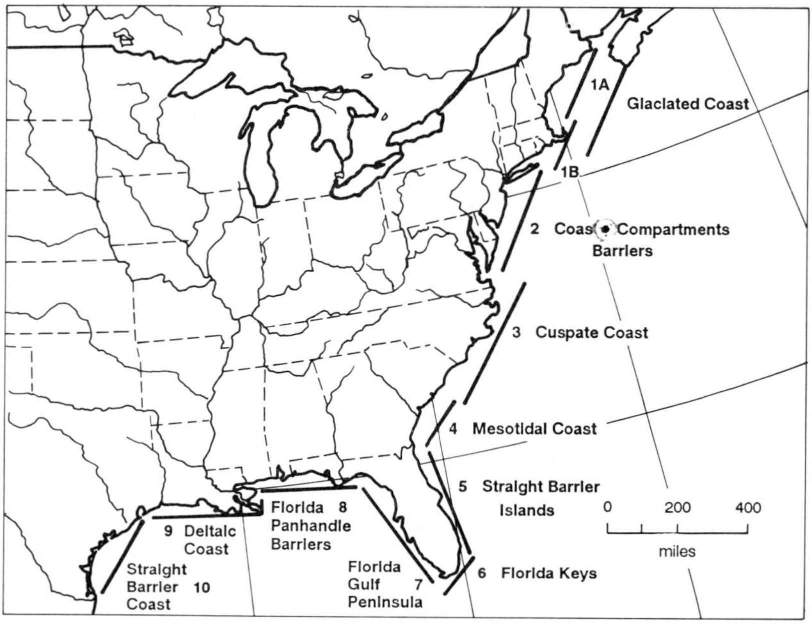

FIGURE 5-1 The Atlantic and Gulf coastlines classified according to geological and morphological criteria. Source: Adapted from Shepard and Wanless (1971).

The cuspate coast, which is best exemplified by Cape Hatteras, North Carolina, extends from Norfolk, Virginia to Cape Romain, South Carolina (Figure 5-1). The "Outer Banks" type barrier islands form a string of sand that protrudes far from the mainland coast along the northern part until merging with the mainland at Myrtle Beach, South Carolina. The historical record shows that the Outer Banks of North Carolina have been breached by many inlets (Swift, 1968).

The mesotidal coast of the Georgia bight extends from southern South Carolina to northern Florida. This physiographic region has the only tide-dominated barrier islands along the U.S. Atlantic and Gulf coastal plains. These islands tend to be relatively short and stubby with marsh-filled lagoons; they often display a "drumstick shape" (Hayes, 1979). The famed Sea Islands (Hoyt and Henry, 1967) are also present in this region. These coastal landforms are distinctive by being composed of a Pleistocene core (often with land surface above the 500-year flood level). Holocene (modern) sediments are plastered onto the seaward face of the old Pleistocene barrier or separated from it by small salt marshes and tidal creeks.

The Sea Islands and mesotidal barriers are more stable than their microtidal counterparts due to the fact that the tidal inlets are located in pre-Holocene drainage channels, and their locations shift within narrowly defined limits (Oertel, 1979). That these inlets are tidally adjusted to better accommodate storm surge flooding, in combination with the antecedent topography and possibly some present day riverine sources of coarse sediment, makes these islands more stable than barriers along the rest of the Atlantic and Gulf coasts.

Straight, narrow barrier islands persist from northern Florida to the Florida Keys, with Cape Canaveral being the only anomaly. The lagoons are narrow and shallow, and the presence of existing inlets, many of which are stabilized, largely precludes future inlet breaching. The coast is low and sandy, with only occasional outcrops of erosion-resistant limestone (Anastasia coquina rock formation) in the beach face.

The Florida Keys are founded on coral reefs, often capped with coral rubble and, in limited areas, faced with sandy beaches produced by the hurricane destruction and wave-abrasion of living or dead corals in the surf zone. Coral reef platforms are more resistant to erosion than barrier islands, but their low elevations make them especially vulnerable to increased flooding and overwash with relative sea level rise. However, coral reefs can continue to grow vertically, which is nature's response to sea level rise, as long as anthropogenic pollution does not kill these ecosystems.

The Florida Gulf peninsula contains a diverse coast of sandy barrier beaches and swamps. The Everglades empties directly into the Florida Bay along a wide front. This physiographic section is largely controlled by Pleistocene limestone rock, which in some places outcrops near sea level to shelter and allow formation of coastal swamps and in others lies tens of feet below the present water surface (Evans et al., 1985). The swampy coast is little developed, but the intermittent, low elevation, microtidal barrier islands have been highly urbanized near such population centers as Tampa.

The Florida panhandle barrier-island system actually extends to the Mississippi delta off Gulfport, Mississippi. The barriers are largely Holocene in origin, except for a Pleistocene core, which constitutes the eastern, bulbous end of Dauphin Island, Alabama. These beach-ridge dominated barrier islands indicate periods of past geological accretion and are presently experiencing strong westward migration due to littoral drift.

The deltaic coast of Louisiana is the most vulnerable to relative sea level rise of the entire U.S. continental area. The sediments are largely fine grained (silts and clays), very organically rich, and subject to compaction (and hence suffer subsidence) and erosion along the shoreline. Much of the Louisiana coast, except for the small area of active delta building, is retreating landward on the order of many meters per year (May et al., 1983). The natural problem of compaction, predominantly a result of loading by sediments deposited in the Mississippi delta region, is greatly compounded by the artificial withdrawal of subsurface fluids. Hence, some areas are sinking at rates of 1 cm/yr or more, drowning salt marshes and pushing the small sandy barriers as thin sand wedges over the adjacent back-barrier sediments. The Chandeleur Islands and Isles Dernieres barrier chains are being fragmented by hurricanes, and it appears that these islands will be lost during the next 100 years even under the present conditions of relative sea level rise (Penland et al., 1985).

The straight barrier coast of Texas has been well studied (McGowen et al., 1977). The small delta of the Brazos River is the only major interruption of these very long, but fairly wide barrier islands. The existing inlets front the mouths of large bays or rivers emptying into the sea (e.g., Brazos and Rio Grande rivers). Generally, the great widths (several miles across) of the Texas barriers, microtidal conditions, and shallow lagoons preclude most inlet activity. Unlike the condition on most of the Atlantic and Gulf coasts, rivers are still providing some coarse-grained material (sand) to the open-ocean coast, so that the barrier islands are not nearly as sand-starved as those found elsewhere. Locally, erosion is a problem, as at Sargents Beach (Herbich, 1975) and as evidenced by the lack of beach in front of portions of the Galveston sea wall.

The Pacific Coast of California, Oregon, and Washington is quite irregular and diverse. There is little to no coastal plain, and cliffs of resistant hard rock or unconsolidated river-fill sediments predominate along this tectonically active Pacific rim. Within small embayments, sandy to gravelly spits can grow, but these landforms are ephemeral, geologically speaking, and are highly unstable features over the short term (Komar, 1976). The geomorphic diversity over short distances largely precludes the typing of this coast into natural units as physiographic regions. Generally, there are three sections.

- From the Mexican border to Point Conception, California, the coast is a nearly continuous, albeit very narrow, beach interrupted by a few headlands.
- From central California to the Columbia River, Oregon, headlands predominate.
- From the Columbia River to the Olympic Peninsula, Washington, the fine-grained, flat-sloped, sandy beaches are composed of river-derived material.

The Alaskan Coast can be roughly divided into four physiographic regions (Shepard and Wanless, 1971):

- fjords of the south coast,
- Aleutian hard rock islands,
- permafrost-dominated lowlands of the west coast, and
- low barriers, spits, and permafrost mainland of the north coast.

In assessing the vulnerability of a coastal area to sea level rise, the best guide is to consider the nature of the sediments (erosion-resistant bedrock or unconsolidated sands, gravels, and clays) and the topography (high to low cliffs versus low sandy barrier spits). Additionally, the degree of existing erosion may serve as an index if future problems result from an accelerated sea level rise.

SANDY COASTLINES

Mean sea level is one of the principal determinants of shoreline position. Swift et al. (1972) suggested that a relationship exists among several factors: sediment supply, wave energy, sea level, and shoreline position. Rising relative sea level tends to cause shoreline recession, except where this trend is offset by an influx of sediment.

The primary reason that a sea level rise would induce beach erosion is that natural beach profiles are concave upward; this geometry results in the wave energy being dissipated in a smaller water volume than without sea level rise, and thus the turbulence generated within the surf zone is greater. The profile responds by conforming to a more gentle nearshore slope, which requires additional sediment to be eroded from the beach.

Most sandy shorelines worldwide have retreated during the past century (Bird, 1976). Progradation has been restricted to coastal areas where excess sediment is supplied by river sources or where the land is being elevated due to tectonic uplift or isostatic glacial rebound. Human interference cannot be considered a primary cause of erosion worldwide since retreat also occurs on sparsely populated and little-developed sandy coasts (Bird, 1976). Such recession could result from an increase in storminess, but this trend would have to be almost worldwide to account for erosion on geographically dispersed sandy shorelines. Therefore, in view of the demonstrated general relative rise of sea level along the U.S. shoreline, the link between shore retreat and sea level rise is based on more than circumstantial evidence; it can be stated that the relationship is causal in nature.

In some areas, it is clear that human actions have caused substantial erosional pressures. Undoubtedly the principal contributor has been the construction of jettied inlets and the deepening of channel entrances for navigation. Along shorelines with high rates of longshore sediment transport, these constructed features trap sediment at the updrift jetty and, if material dredged from the navigation channel is not placed on the downdrift beaches, cause an amount of downdrift erosion equal to the reduction in transport. At some Florida entrances, tens of millions of cubic yards of dredged material have been placed in water depths outside the littoral system. This has resulted in very high erosion concentrated downdrift of the entrances. Some of these shorelines were stable or accreting in their natural condition, prior to inlet modification.

Geologic Indicators

The geologic record of the Atlantic and Gulf coastal plains over the last 10 million years indicates that sea levels have fluctuated by 200 m or so during this time period. Five distinct transgressive coastal systems have been identified on the Delmarva peninsula from geomorphic and subsurface data. Each was produced during interglacial high sea levels and range in age from more than 1 million years to 60,000 years (Demarest and Leatherman, 1985). Sedimentological and historical evidence for four minor transgressive phases or pulses, with sea level fluctuations of less than 1 m, during the last 2,000 years have been found along the Friesland barrier islands in the Netherlands (Bakker, 1981). The modern transgressive pulse during the overall Holocene transgression presumably began in the eighteenth century, when history indicates an increase in storm surge damage and coastal flooding.

Long periods of sedimentary accretion resulting in beach ridges have been arrested or the trend reversed during the past century. Teichert (1947) reported that beach ridge formation ceased slightly more than 100 years ago, and the western Australian coralline shore is now subject to erosion by the sea. In Nigeria, Pugh (1954) noted that earlier progradation had similarly given way to retrogradation on sandy shorelines. Bogue Banks, along the Outer Banks of North Carolina, is a barrier island composed of parallel sets of beach ridges, which have prograded seaward during the past 3,000–4,000 years, but now the beaches are narrow and dunes are actively wave cut during annual winter storms (Steele, 1980). Similar reversals in trend, from long-term accretion to recession, have been noted by many investigators working along sedimentary coasts in the United States (e.g., Tanner and Stapor, 1971) and worldwide (Davies, 1957).

Onset of the present transgressive pulse, attested to by marked beach and dune erosion, has varied geographically depending upon local differences in sand supply and wave energy. Information from 73 correspondents in 39 coastal countries showed that less than 10 percent of the length of the world's sandy shorelines have prograded, more than 60 percent have retrograded, and the balance have been relatively stable or have shown no consistent trend during the past century (Bird, 1976).

Other geologic indicators of shore retreat are wave-cut cliffs, which occur worldwide (Sunamura, 1983). Exhumation of salt marsh peat on beach faces indicates upward and landward barrier migration. Most barrier island coasts have been retreating for at least the past few hundred years, as clearly indicated by these back-barrier peat outcrops, exposed on the lower beach foreshore after severe storms. Peat outcrops have been reported in widely dispersed areas along the U.S. Atlantic and Gulf coasts, including Nauset Spit, Massachusetts (Leatherman, 1979b); coastal Delaware (Kraft, 1971); Assateague Island, Maryland (Leatherman, 1979a); Cape Hatteras, North Carolina (Swift, 1968); and Sargent Beach, Texas (Herbich, 1975).

TABLE 5-1 National Assessment of Shore Erosion (miles)

Location	Total Shoreline	Erosional	Nonerosional
North Atlantic	8,620	7,460	1,160
South Atlantic Gulf	14,620	2,820	11,800
Lower Mississippi	1,940	1,580	360
Texas Gulf	2,500	360	2,140
Great Lakes	3,680	1,260	2,420
California	1,810	1,550	260
North Pacific	2,840	260	2,580
Alaska	47,300	5,100	42,200
Hawaii	930	110	820
Total	84,240	20,500	63,740

SOURCE: U.S. Army Corps of Engineers (1971).

Historical Records

Historical records also indicate the prevalence of shore recession during at least the past century. The National Shoreline Study by the U.S. Army Corps of Engineers (1971) was the first overall national appraisal of shore erosion problems. Of the over 84,000 miles of United States ocean and Great Lakes shorelines, significant erosion occurs along 20,500 miles or 25 percent of the total (Table 5-1). Excluding Alaska, it shows that 43 percent of the shoreline is undergoing significant erosion. It should also be noted that a significant portion of the shoreline is categorized by the U.S. Army Corps of Engineers (1971) as noncritical, which does not connote nonserious. In these cases the problems appear to be amenable to land use controls and other management techniques rather than relying upon engineering measures to halt erosion.

More recently, May et al. (1983) have assembled data derived from aerial photography of the U.S. Geological Survey dating back to the late 1930s, providing a maximum record of 40–50 years. The National Ocean Service of the National Oceanic and Atmospheric Administration (NOAA) also has made efforts to determine the historical rate of shoreline change along portions of the U.S. continental coast using historical maps and charts (NOS "T" sheets). The data base includes most of the mid-Atlantic Coast as well as South Carolina and parts of California. These data allow for the quantification of historical shoreline changes over 100–150 years of record. The NOS maps show a general pattern of pervasive shore recession except for local anomalies (Everts et al., 1983).

The existing data sets (Table 5-2) have been grouped by state for comparative purposes (May et al., 1983). The national average (unweighted) shoreline erosion rate is 0.4 m/yr. Along the Atlantic Coast, the average erosion rate is about 0.8 m/yr with the Virginia barrier islands exhibiting the highest rates of erosion (Leatherman et al., 1982). The Gulf Coast states are distinguished by the highest average erosion rate in the nation (1.8 m/yr). The deltaic coast of Louisiana is by far the most dynamic (4.2 m/yr erosion; May et al., 1983). The Pacific coastline is essentially stable, although more than half of the shore is hard rock. The erosion rate can be tabulated by landform type for comparative purposes (Table 5-3). Table 5-3 may be useful in ascertaining the shoreline erosion rate for a site-specific area, such as along the geomorphically diverse Pacific Coast.

Techniques of Projecting Shoreline Retreat Due to Sea Level Rise

Rising sea level is accompanied by a general recession of the shoreline due to inundation or erosion. Inundation is the submergence of the otherwise unaltered shore, while erosion is the physical removal of beach material. Direct submergence of the land occurs continuously through time and is particularly evident in coastal bays where upland is slowly converted to coastal marshlands. Submergence, however, accounts for only a small portion of the net shore recession along exposed, sedimentary coasts (Hands, 1976).

Several different approaches can be used to model the resulting shoreline configuration as a function of sea level rise. The simplest method uses the drowned-valley concept (Figure 5-2), in which preexisting topography along shorelines is considered fixed and combined with increased sea level to project new shorelines (Kana et al., 1984). Slope is the controlling variable: steep-sloped areas will experience little horizontal shoreline displacement with each increment of water level rise, while gently sloping shores will undergo a much broader area of flooding for a given sea level rise. This is the preferred methodology for immobile substrates, such as

TABLE 5-2 Shoreline Erosion Rate Based on Historical Aerial Photographs by State and Region

Region	Average Shoreline Change Rate (m/yr)[a]	Standard Deviation of Shoreline Change Rate (m/yr)	Extreme Shoreline Change Rates (m/yr)[a] Maximum Accretion	Maximum Erosion	Number of Sample Data Points[b]
Atlantic Coast	-0.8	3.2	25.5	-24.6	510
Maine	-0.4	0.6	1.9	-0.5	16
New Hampshire	-0.5	--	-0.5	-0.5	4
Massachusetts	-0.9	1.9	4.5	-4.5	48
Rhode Island	-0.5	0.1	-0.3	-0.7	17
New York	0.1	3.2	18.8	-2.2	42
New Jersey	-1.0	5.4	25.5	-15.0	39
Delaware	0.1	2.4	5.0	-2.3	7
Maryland	-1.5	3.0	1.3	-8.8	9
Virginia	-4.2	5.5	0.9	-24.6	34
North Carolina	-0.6	2.1	9.4	-6.0	101
South Carolina	-2.0	3.8	5.9	-17.7	57
Georgia	0.7	2.8	5.0	-4.0	31
Florida	-0.1	1.2	5.0	-2.9	105
Gulf of Mexico	-1.8	2.7	8.8	-15.3	358
Florida	-0.4	1.6	8.8	-4.5	118
Alabama	-1.1	0.6	0.8	-3.1	16
Mississippi	-0.6	2.0	0.6	-6.4	12
Louisiana	-4.2	3.3	3.4	-15.3	106
Texas	-1.2	1.4	0.8	-5.0	106
Pacific Coast	-0.0	1.5	10.0	-5.0	305
California	-0.1	1.3	10.0	-4.2	164
Oregon	0.1	1.1	5.0	-5.0	86
Washington	0.5	2.2	5.0	-3.9	46
Alaska	-2.4	2.0	2.9	-6.0	69

[a] Negative values indicate erosion and positive values indicate accretion.
[b] Total number of 3-min grid cells over which statistics are calculated.

SOURCE: May et al. (1983).

rocky or armored shorelines, or where the wave climate is subdued, as on the sheltered coasts of embayments.

Several approaches to shoreline recession that have been employed to date are largely based on the erosional potential of sea level rise: (1) extrapolation of historical trends (Leatherman, 1984b), (2) the Bruun rule (Hands, 1981, Weggel, 1979; Bruun, 1962), (3) the sediment budget method (Everts, 1985), and (4) the dynamic equilibrium model (Dean, 1983). These methodologies, including applications and limitations, will be discussed in the order outlined.

TABLE 5-3 Historical Shoreline Erosion Rate According to Coastal Landform Type

Region	Average Shoreline Change Rate (m/yr)[a]	Standard Deviation of Shoreline Change Rate (m/yr)[a]	Extreme Shoreline Change Rates (m/yr)[a] Maximum Accretion	Extreme Shoreline Change Rates (m/yr)[a] Maximum Erosion	Number of Sample Data Points[b]
Mud flats					
Fla.	-0.3	0.9	1.5	-1.5	9
La.-Tex.	-2.1	2.2	3.4	-8.1	84
All Gulf	-1.9	2.2	3.4	-8.1	93
Rock shorelines					
Atlantic	1.0	1.2	1.9	-4.5	36
Pacific	-0.5	--	-0.5	-0.5	7
Pocket beaches					
Atlantic	-0.5	--	-0.5	-0.5	9
Pacific	-0.2	1.1	5.0	-1.1	144
Sand beaches					
Maine-Mass.	-0.7	0.5	-0.5	-2.5	17
Mass.-N.J.	1.3	1.3	2.0	4.5	22
Atlantic	-1.0	1.0	2.0	-4.5	39
Gulf	-0.4	1.6	8.8	-4.5	121
Pacific	-0.3	1.0	0.7	-4.2	19
Sand beaches with rock headland	0.3	1.9	10.0	-5.0	134
Deltas	-2.5	3.5	8.8	15.3	155
Barrier islands					
La.-Tex.	-0.8	1.2	0.8	-3.5	76
Fla.-La.	-0.5	1.7	8.8	-4.5	82
Gulf	-0.6	1.5	8.8	-4.3	158
Maine-N.Y.	0.3	2.6	4.5	-1.5	12
N.Y.-N.C.	-1.5	4.5	25.5	-24.6	153
N.C.-Fla.	-0.4	2.6	9.5	-17.7	256
Atlantic	-0.8	3.4	25.5	-24.6	421

[a]Negative values indicate erosion and positive values indicate accretion.
[b]Total number of 3-min grid cells over which the statistics are calculated.

SOURCE: May et al. (1983).

Historical Trend Analysis

Trend analysis is essentially a calibration procedure using historical shoreline data. Shoreline response is based on the historical trend with respect to local sea level change during a given time period. This procedure accounts for the inherent variability in shoreline response based on differing coastal processes, sedimentary environments, and coastline exposures.

The method of projecting shoreline movement due to accelerated sea level rise is as follows (Leatherman, 1984b):

1. Observe historical shoreline movement for as long a period of record as possible using quantitative data from accurate maps, charts, and vertical aerial photographs.

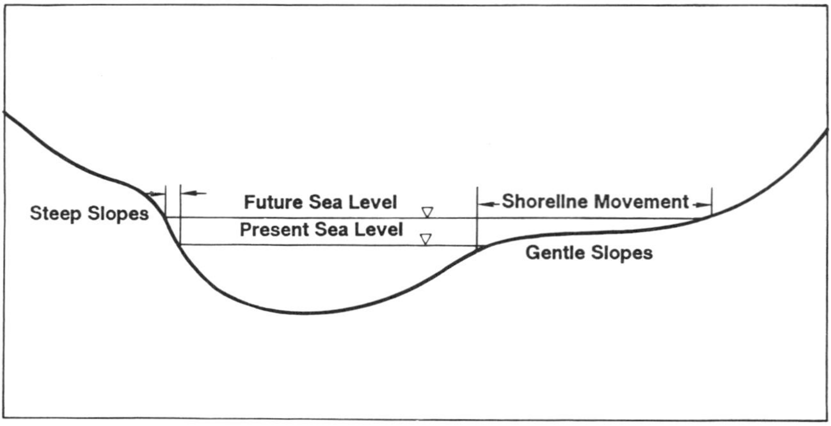

FIGURE 5-2 Schematic cross section of drowned-valley concept of sea level rise. Note that the shoreline movement greatly depends on the land slope. Source: Adapted from Kana et al. (1984).

2. Establish a centimeter (foot) per year relationship for different shoreline types and wave exposures, using the historical rate of sea level rise for that area (Hicks et al., 1983).

3. Develop a hypothesis or rule of thumb on the basis of which to project further relationships between sea level rise and shoreline movement. It is assumed that the amount of recession from the historical record is directly correlated with the rise rate of sea level. Therefore, a threefold rise in sea level will result in a threefold increase in the recession rate, assuming lag effects in shoreline responses are small compared to the overall accuracy of extrapolation.

Tide gauge records document the local rate of sea level change over the period of record. Shoreline charts of the NOS, formerly the U.S. Coast Survey and U.S. Coast and Geodetic Survey, are used for shoreline comparisons. The NOS "T" sheets were made from field surveys and are presently the most accurate maps of the shoreline (Shalowitz, 1964).

This type of analysis can be undertaken for any coastal plain shoreline. The easily eroded, unconsolidated sediments and gently sloping, low-lying topography make the projections straightforward, except where modified by coastal engineering structures. The underlying assumption of this analysis is that shorelines will respond in similar ways in the future, since sea level rise is the predominant driving function and all other parameters remain essentially constant. The methodology has been applied to Galveston, Texas and Ocean City, Maryland to estimate future shoreline changes with accelerated sea level rise.

The Bruun Rule

Bruun (1962) was the first to formulate the relationship between rising sea level and the rate of shoreline erosion. Bruun's argument is based largely on the concept of an equilibrium beach profile, which has had a long history dating back to Fenneman (1902). The term "equilibrium profile" is a statistical average profile that maintains its form apart from small fluctuations, including seasonal effects at a particular water level (Bruun, 1954). Use of the term "equilibrium" in this context is not inconsistent with the recognition of seasonal, storm, or other temporal profile fluctuations.

Bruun's (1962) quantitative relationship for the equilibrium profile can be expressed in the form

$$h = \mathrm{A}x^{2/3},$$

where h is the water depth, x is the horizontal distance from shore, and A is the constant for each profile. The Bruun rule provides for a profile of equilibrium in that the volume of material removed during shoreline retreat is transferred onto the adjacent inner shelf, thus maintaining the original beach profile and nearshore shallow-water conditions (only further inland). Figure 5-3 depicts this two-dimensional approach of sediment balancing between eroded and deposited quantities in an on/offshore direction. With an incremental rise in sea level, it is clear that additional sand must be added to the below-water portion of the beach profile; assuming no longshore variations, this sand must be derived from beach erosion. The so-called "Bruun rule" can be stated as

$$R = \frac{SWG\mathrm{a}}{h_*}$$

in a modified form presented by Hands (1981, 1976) in which R represents shoreline recession, S is sea level rise, W is the width of the "active" portion of the profile participating in the adjustment, and h_* is the vertical distance over which the adjustment

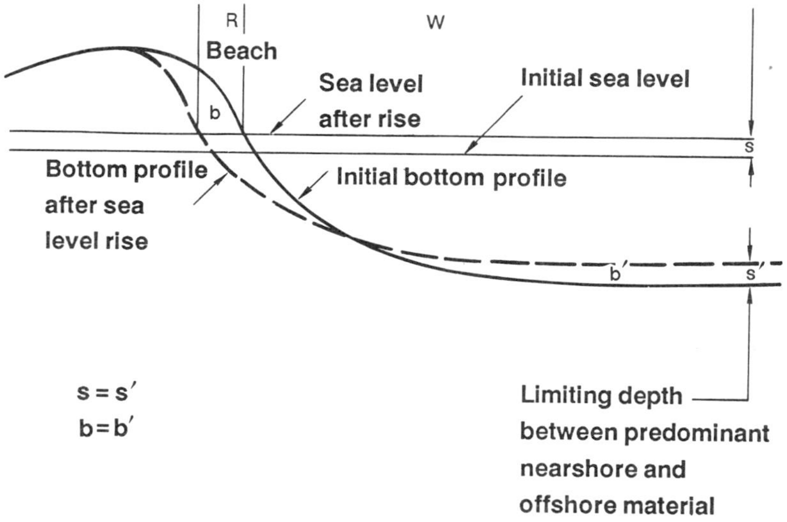

FIGURE 5-3 The Bruun rule: a rise in sea level causes beach erosion. If the sea rises 1 ft, so will the offshore bottom. The sand necessary to raise the bottom (area b′) can be supplied by artificial beach nourishment or by waves eroding the upper part of the beach (area b). Source: Adapted from Schwartz (1967).

occurs, including the above-water and below-water portions of the adjusting profile (see Figure 5-3). The factor *G*a is the overfill ratio, which quantifies the amount of material to be placed on the beach to yield a unit volume of compatible beach sand. This factor allows the composition of the eroding beach or bluff material to be included.

Bruun (1962) found reasonable agreement between the predicted and actual erosion rates along the southeast coast of Florida. In general, it was found that a rapid rise in sea level of S causes a shoreline recession of about $100S$, which translates to about 1 m/yr for his study area. Schwartz (1965) used small-scale wave tank tests to verify this hypothesis.

Bruun's concept is straightforward and intuitively appealing, but it is difficult to confirm or quantify without precise bathymetric surveys and integration of complex nearshore profiles over long periods of time. Also, definition of the active profile boundaries in the seaward direction necessitates the selection of a pinch-out depth of significant sediment motion, a rather vague concept in an oceanic wave environment. The problem is compounded when an attempt is made to quantify a relatively confined zone of erosion (e.g., the narrow upper shoreface/beach/dune zone) with a broad zone (the lower shoreface/inner shelf) over which eroding sediment can be thinly spread.

Hallermeier (1981) has shown that the so-called "depth of closure" or "pinch-out depth" varies considerably around U.S. shorelines. Dietz's (1963) 9-m wave base is adequate as a first approximation, but this average depth cannot be applied to specific areas to obtain quantitative results. Hallermeier concluded that the appropriate depth value depends upon its application and is related to some specific nearshore wave-height statistic. In the case of sediment budget calculations with time spans on the order of decades, typical values of the depth of closure along the U.S. continental coasts range from 4 to 8 m.

The Great Lakes serve as a natural laboratory for documenting the effects of rising water levels on shore position. Hands (1976) has carried out a field evaluation to assess the applicability of the Bruun rule. Beach and nearshore changes were monitored at 25 profiles along a 50-km stretch of Lake Michigan over a 9-year period of persistently rising and then stable annual mean lake levels. Due to climatic periods of wet and dry conditions, lake levels have fluctuated by as much as 1.8 m (6 ft) in little over a decade. During 1969, lake levels were again approaching a high stage, resulting in significant erosion of sandy beaches and cliffs along many lake shores. Because the Great Lakes are not subject to astronomical tides to any appreciable degree and are not influenced by hurricanes or long-period swell (Hands, 1983), these complicating variables were eliminated.

Hands (1976) found that the Bruun rule was well satisfied in the field surveys of beach and nearshore profiles during rising lake levels. The volume of sand eroded from the beach nearly matched offshore deposition, providing the first actual field verification of this hypothesis. Hands (1976) also found that deposition extended offshore to a distance where the water depth is equal to roughly twice the wave height of a 5-year storm.

Profile retreat was found to lag behind the lake level rise. Rising water levels establish a potential for erosion, but realization of the potential requires sediment redistribution, that is, work that depends on energy being available.

It may be useful to view the allied roles of sea level rise and wave energy by considering sea level changes as setting the stage for profile adjustments by coastal storms. Long-term sea level rise places the beach/nearshore profile out of equilibrium, and sporadic storms accomplish the geologic work in increments, each depending on the magnitude and duration of the associated storm. Major storms are required to mobilize the bottom sands at great depths offshore and thus fully adjust the profile to the existing water level position.

Therefore, the underlying assumption is that beach equilibrium will be the result of water level position in a particular wave climate setting. Shoreline response lag times are tied to storm intensity and frequency, as shown by Hands (1976). The lag in shoreline response to lake level was shown to be rather short (about 3 years). This rapid response time is due to the fact that the Great Lakes are subject to frequent storm activity in the fall and winter before surface icing.

With some qualifications, the Great Lakes research may prove to be a useful analog in considering the response of open-ocean shores to long-term sea level rise. For example, the mid-Atlantic Coast is subject to both extratropical (northeasters) and tropical (hurricanes) storms, both of which generate large waves capable of significant beach erosion. However, there has been a lull in major storm activity along the Atlantic Coast; Hurricane Donna in 1960 and the "Ash Wednesday" northeaster of 1962 were the last major storms of record. Therefore, areas such as Ocean City, Maryland are probably considerably out of adjustment with sea level change (Leatherman, 1985). An appreciable time lag in shoreline response, depending upon local storm frequency, can only be dealt with in a statistical manner.

Application of the Bruun rule also depends on local conditions. For example, Bruun (1983) provides four different situations: (1) closed basin, (2) wide shelf, (3) narrow shelf, and (4) profile with deposit slope. In a closed basin such as a lake, a restricted fetch and often shallow depth would limit incident wave energy and closure depth. Wide shelves would be represented by trailing edge or marginal coasts, such as the Atlantic and Gulf coasts, respectively; the U.S. Pacific Coast shelf is quite narrow. Rosen (1978) has shown that there is a nearshore platform rimming the shore of the Chesapeake Bay and that sand deposition from shore erosion occurs shoreward of the 3.6-m contour, defining this geologically wave-beveled surface.

Dean and Maurmeyer (1983) have generalized the Bruun rule

to represent the case of landward and upward-migrating barrier island systems. The active profile of change must be expanded to include barrier dimensions and overwash and inlet transport into the adjacent lagoon. In this case the barrier island unit, including the seaward and bayward active profile segments, is considered to move landward and upward without change in form. The vertical movement keeps pace with sea level rise and the landward movement is of such a magnitude to conserve sediment. If there were no long-term averaged landward transfer of sediment to maintain barrier width (Leatherman, 1979a), the shore recession accompanying sea level rise would cause the island to erode away literally and "drown in place." Since overwash and inlet sand are being lost to the beach and nearshore profile, this generalized approach would always predict greater retreat than does the Bruun rule.

Shoreline adjustment along straight sandy shores that are exposed to ocean waves due to a rise in mean sea level is projected under the assumption that the beach profile is retained, but it moves upward by the amount of the rise and landward by the distance required to supply sand to fill out the profile. The visible beach is reduced in width, but the volume of sand removed from the beach lies seaward.

On beaches where there are both longshore sand transport and fixed points such as rocky headlands or other obstructions that may fix the position of the beach face, the local effects of a rise in mean sea level may be altered by the temporary accumulation of sand updrift of the fixed position and denial of this volume to the beach downdrift. The distance updrift and downdrift to which this effect extends will depend upon the configuration of the fixed position and the dominant wave direction. Under such circumstances, the magnitude predicted by the effect of the Bruun rule may be lessened updrift, and augmented downdrift, of such obstructions.

Sediment Budget Approach

The sediment budget approach is a method of quantifying sources and sinks for a given control volume as detailed in the *Shore Protection Manual* (U.S. Army Corps of Engineers, 1984), and is essentially a formulation of the conservation of volume. Although the approach is straightforward in concept, its application requires accurate data to yield valid results; data needs include annualized values of littoral drift, inlet losses, overwash, and offshore leakage. Quantification of nearshore and shoreface profile changes is particularly problematic, since small vertical changes over such broad areas can represent huge volumetric amounts. Field measurements are probably the least reliable in this zone because they are generally obtained by boat and fathometer.

There are also uncertainties in characterizing the inner shelf as a source or sink. Contributions to the offshore from the beach and shoreface are substantial; Hayes (1967) and Nummedal and Snedden (1987) showed that Hurricane Carla carried fine sands as far as 50 km offshore of Padre Island, Texas into water depths of more than 50 m. In other areas, such as along the south shore of Long Island, New York, it appears that sand is moving onshore from the inner shelf to augment the longshore transport systems. Erosion of the updrift headland area (Montauk Point to Southampton, New York) can be shown to account for only half of the material in transport as littoral drift (Taney, 1961).

Everts (1985) estimated losses to the shoreface (offshore leakage in excess of equilibrium profile) by considering nearshore shoals as sand losses to the landward-retreating barrier. Field and Duane (1976) have shown that these linear sand ridges are dynamic, rather than relict features, and are believed to be initiated on the shoreface. With continued sea level rise and concomitant landward barrier retreat, these large sand bodies are essentially left behind, albeit reworked surficially. Eventually they become detached from the nearshore sand-sharing system and represent a net loss of sediment to the barrier system.

Swift (1975) has shown that nearshore ridges become larger in an offshore direction, suggesting their continued growth through time. This additional sand, however, may not be derived from the nearshore zone, but instead supplied by local sources as ridges become higher and swales deeper (Goldberg et al., 1979). Other losses or gains, such as overwash and aeolian transport, beach nourishment, and sand mining, can be reasonably well determined from past geological studies and public records.

Everts et al. (1983) have applied a sediment budget model to several coastal areas including Smith Island, Virginia, where they estimated that sea level rise accounts for only 14 percent of the shoreline recession. This barrier has experienced rapid shoreline recession, averaging 5.6 m/yr largely due to insufficient supplies of updrift littoral drift. For the Outer Banks of North Carolina, Everts et al. (1983) attribute 73 percent of the shore recession to sea level rise by using best estimates of the input variables from the scientific literature.

There are major differences in sediment budgets from site to site, and each area must be evaluated individually with respect to the existing sediment budget and the effects of present and future sea level rise.

Dynamic Equilibrium Model

The dynamic equilibrium model attempts to account for the transient response characteristics of a beach profile due to changes in the forcing function (i.e., changing water level and wave conditions). The transient response is most important for severe storm conditions; for example, the water level may fluctuate 3–6 m in a period of 6–12 hours during a hurricane.

Dean (1977) investigated the concept of the equilibrium beach profile. As previously indicated, Bruun (1954) developed the following empirical equation between water depth h and distance x from the shoreline:

$$h = Ax^{2/3},$$

where A is a shape factor, depending on stability characteristics of the bed material.

Dean (1977) analyzed 502 beach profiles from the U.S. Atlantic and Gulf coasts to show that the exponent's value (0.67) was indeed correct on an average basis. The coastline was segmented into geomorphic regions, and there appears to be a geographic trend to the data. The parameter A was found to be a function of sediment (and possibly wave) characteristics, since steeper profiles were associated with coarser sand, low wave height, and long wave periods.

The monotonic equilibrium profile of the form $h = Ax^{2/3}$ is consistent with a uniform wave energy dissipation per unit water volume within the surf zone (Dean, 1977). It is also known that beaches respond to increases in water level by erosion of sediment in shallow water and by deposition of this sediment in deeper water (Moore, 1982; Hands, 1981). Therefore, Dean (1983) proposed that offshore sediment transport Q_s, per unit width could be expressed by

$$Q_s = k(D - D_*),$$

where k is the rate constant ($2.2 \times 10^{-6} m^4/N$), D is the wave energy dissipation rate per unit volume, and D_* is the equilibrium wave energy dissipation rate per unit volume. The units of D and D_* are $N/m^2/S$.

These equilibrium beach profile concepts, along with the continuity equation, form the basis of a two-dimensional, numerical erosion model. The cross-shore transport of sand is cast in a finite difference form, and the time-varying water level and wave height conditions are prescribed. A numerical solution yields the time-dependent beach and dune response during a storm. Realistic analyses can be based on a probabilistic model that properly represents the storm statistics. Verification studies, using Hurricane Eloise erosion field data, show that the numerical analysis is subject to probable errors of ±25 percent; some of this discrepancy may be due to omission of longshore transport or overwash effects (Kriebel and Dean, 1985).

The numerical model by Kriebel and Dean (1985) has been applied to Ocean City, Maryland to forecast future rates of beach erosion with sea level rise. Corrections to account for other sand volume losses (e.g., overwash and aeolian transport, littoral drift) were taken from Everts (1985) for application to this cross-shore transport model. Application of the dynamic equilibrium model with the sediment budget overlay indicated that the existing rate of sea level rise accounts for about 20 percent of the historical shoreline retreat rate for Ocean City, Maryland (1.9 ft/yr, Leatherman, 1985). Due to the nonlinear erosion response to accelerated sea level rise, projected shore erosion rates also accelerate in the future (Kriebel and Dean, 1985). In the absence of landward barrier migration or some human intervention, Ocean City will eventually drown in place if long-term rates of sea level rise are realized.

For comparison, projected erosion at Ocean City, Maryland for various rates of sea level rise have been tabulated for the four methodologies: Bruun, Everts, Leatherman, and Kriebel/Dean (Table 5-4). Although the methodologies produce substantially different results, all are consistent in predicting more rapid retreat

TABLE 5-4 Projected Erosion at Ocean City, Maryland in Meters (ft) of Shoreline Retreat Relative to Its Current Position

Methodology	2000	2025	2050	2075
	Current Trends			
Bruun[a]	5 (16)	11 (36)	17 (57)	23 (75)
Everts	21 (68)	47 (153)	73 (238)	99 (323)
Leatherman	12 (39)	26 (85)	41 (134)	56 (182)
Kriebel/Dean	20 (66)	47 (153)	70 (231)	95 (102)
	Mid-range Low			
Bruun[a]	7 (22)	22 (72)	43 (140)	70 (231)
Bruun adjusted[b]	23 (74)	58 (189)	98 (322)	147 (482)
Everts	26 (84)	73 (238)	132 (434)	215 (205)
Leatherman	20 (64)	56 (182)	105 (345)	174 (571)
Kriebel/Dean	NC	55 (180)	NC	140 (460)
	Mid-range High			
Bruun[a]	12 (38)	32 (106)	63 (206)	105 (346)
Bruun adjusted[b]	27 (90)	68 (223)	118 (388)	181 (592)
Everts	29 (95)	83 (273)	156 (511)	268 (878)
Leatherman	27 (89)	76 (250)	147 (483)	249 (812)
Kriebel/Dean	NC	66 (216)	NC	168 (550)

NOTE: NC = not calculated.

[a]Bruun rule includes only the impacts of sea level rise.
[b]Bruun rule adjusted includes 2.6 ft/yr due to factors other than sea level rise. Because 2.6 ft/yr is derived from Everts, Bruun adjusted is equal to Everts for current trends.

SOURCE: Titus (1985).

rates near the end of the time span considered and substantial potential shoreline changes within the entire time span. The longshore losses along this portion of the coastline are believed to be due to the presence of a nodal point located at South Bethany Beach (U.S. Army Corps of Engineers, 1980).

BLUFF AND CLIFF RETREAT

While most of the attention by coastal geomorphologists and engineers has been directed at studying sandy beaches, cliff retreat is a significant problem along large portions of the nation's coast (i.e., the Pacific Coast, the Great Lakes, and parts of the New England and New York coasts). Increases in water level will only accelerate the erosion rate as has been clearly shown by Hands (1981) along the Lake Michigan shore.

Elsewhere, the high cliffs of unconsolidated sands and gravel along outer Cape Cod, Massachusetts are eroding at an average rate of 2.2 ft/yr based on more than 100 years of field survey data. Dalrymple et al. (1986) indicate that bluff recession in Chesapeake Bay is related to the heights of the bluffs and their compositions, as well as the available wave energy.

Kuhn and Shepard (1981) showed that the unconsolidated sedimentary cliffs of southern California recede in an episodic manner, corresponding to rainfall and storm wave attack during unusually severe winter storms. Thornton et al. (1985) derived an empirical relation between surge level and wave runup and cliff retreat based on studies of Monterey Bay, California.

As previously mentioned, cliffs of crystalline rock are essentially stable with response times to sea level rise much longer than those of sandy shorelines. Thus, for parts of the Pacific Coast and almost all of the rocky Maine coast, cliff position is unchanged over historical periods of record. Sunamura (1983) provides a review of cliff erosion processes.

TIDAL INLETS

Along the barrier island coastlines of the United States, inlets provide hydraulic connections between the back-barrier environments and the ocean. In their natural conditions, inlets can migrate along the shoreline, whereas when stabilized by jetties, they are fixed in position to provide reliable navigation channels.

An inlet can be characterized by its tidal prism, the total flow of water through the inlet from low to high tide, and the amount of sand moving locally in the littoral transport system. Inlets with small tidal prisms have little ability to scour and erode sand transported into the inlet from the adjacent shores. Often these inlets have very pronounced ebb tidal deltas, shallow enough to permit waves to move sand past the inlet. Inlets with reduced sediment transport environments or large tidal prisms have ebb tidal deltas located in deeper water or farther out to sea. In either case the amount of sand capable of bypassing an inlet modified for improved navigation is very much less, and severe downdrift erosion can result (Bruun and Gerritsen, 1960).

The ebb tidal delta and the flood tidal delta in the backbay consist of platforms (swash-platforms) on the ebb tidal delta or ramps (flood ramps), which are separated by channels kept clear by the tidal currents.

The size of the ebb tidal delta is roughly proportional to the tidal prism (Walton and Adams, 1976), and often represents the trapping of millions of cubic meters of sand unavailable to the neighboring beaches. The development of new inlets or the stabilization of existing inlets generally results in the development of large or larger ebb tidal deltas, impounding greater amounts of sand, thereby reducing the sand available to the beaches. Artificial sand bypassing, consisting of a floating or land-based dredge pump which discharges sand onto a downdrift beach, is used at several inlets (e.g., Lake Worth Entrance, South Lake Worth Inlet, and Hillsboro Inlet, Florida) to augment any natural bypassing of sand; however, as a general rule a tidal inlet represents a sink of beach sand. A recent study by the state of Florida (1986) showed that most of the state's eroding areas were next to tidal inlets, implying that effective bypassing of sand at the inlets would reduce many of the erosion problems.

The basic effects of sea level change on tidal prism and inlet cross-sectional area were discussed in general terms earlier in this report. The magnitude of change in tidal prism in response to sea level rise is highly dependent on conditions along the bay shoreline. Bays surrounded by Pleistocene uplands generally have relatively steep shorelines, so that rising sea level will have only a minor impact on changes in tidal prism. For example, an estimated 1 m rise in local sea level at Indian River Inlet, Delaware may cause only a 2 percent change in prism of Indian River Inlet.

Shallow bays surrounded by extensive wetlands will expand rapidly in response to a rise both because of the gentle slope and the deterioration of the marshes in response to water level increases. Barataria Bay, Louisiana has increased its surface area about 10–15 percent over the last century in response to about 1 m of local relative sea level rise in that area.

Of perhaps greater importance is the change in sand storage volume of the ebb and flood deltas. If the prism increases, there is likely to be a corresponding increase in the volume of these shoals. Furthermore, as the sea level rises the deltas must grow in elevation to keep up with the rise, implying that any natural bypassing of sand will be reduced and that downdrift erosion will increase.

Stabilized inlets will be affected strongly by a large sea level rise. The protective jetties, which retard the ability of the littoral drift to enter the navigational channel and reduce the wave climate in the channel, will become less effective as they are submerged. Also, the stability of the jetties is reduced due to the aforementioned greater wave heights as a result of sea level rise.

WETLANDS

Wetlands account for most of the land less than 1 m above sea level. These extensive marshes, swamps, and mangrove forests fringe most of the U.S. coastline, particularly along the Atlantic and Gulf coasts. Coastal wetlands serve as nurseries for fish and shrimp, many birds, and fur-bearing animals. They are vital to coastal recreation, to the maintenance of water quality, and as a buffer against shore erosion.

Their estimated original extent in the United States was 5 million acres or about 20,200 km^2 (7,800 mi^2) (Hoese, 1967). This acreage has been significantly reduced through a variety of actions including an early widespread practice of filling marshlands in urban areas. Wetlands loss has also been caused by other human actions, such as the construction of canals and waterways and the diversion of fluvial sediment to the offshore.

In response to this loss, several federal and state programs have been designed to prevent wetlands destruction. Specifically, Section 10 of the federal Rivers and Harbors Act, Section 404 of the federal Clean Water Act, and Executive Order 11988 on floodplain management all establish permit requirements for actions affecting waterways and wetlands. In general, the wetlands policy of both the U.S. Army Corps of Engineers and the Environmental Protection Agency (EPA) is to discourage issuance of a permit for an activity that would involve alteration of wetlands. However, the effectiveness of this permitting process has been questioned. The congressional Office of Technology Assessment (OTA) concluded that permit applications for wetlands alteration are still rarely denied (OTA, 1984). The continuing human destruction of wetlands should be kept in mind for the proper perspective when considering sea level rise and its potential effects on wetlands deterioration.

Ecological conditions in coastal marshes range from marine to nearly terrestrial. A change in controlling factors, such as water salinity or tidal and wave energy, will cause a displacement in marsh zonation. Generally, coastal marshes are divided into low and high marsh based on their elevation relative to sea level (Redfield, 1972). Since marsh plants are attuned to particular mean water levels (e.g., *Spartina patens*, salt meadow grass, grows at mean high tide), a rise in sea level will shift the distribution of plant species proportionally landward. Beyond this fundamental response to variation in relative sea level, however, a more complex set of attendant responses may occur, tied to the type of marsh considered. Thus, anticipated changes in coastal marshes must be assessed within the context of the basic marsh types that characterize U.S. coasts.

Marshes have been classified on the basis of the flora present (Redfield, 1972) and salinity and floristic relations (Chabreck, 1972), and functionally on the basis of geologic/geomorphic processes (Stevenson et al., 1986). Nevertheless, with respect to the future effects of a rise in sea level, coastal marshes may be broadly divided into back-barrier marshes, estuarine (brackish) marshes, and tidal freshwater marshes.

Back-Barrier Marshes

Back-barrier marshes occur along the bay sides of barrier systems of the Atlantic and Gulf coasts. Studies (e.g., Zaremba and Leatherman, 1986) show that these marshes are formed and destroyed rapidly in such dynamic environments. Maintenance of these marshes is therefore more a function of barrier stability than the pace of upward growth of the marsh surface, since sediment supplies are ample (Letzsch and Frey, 1980). For barriers rapidly migrating landward, there may be a net decline in back-barrier marshes. This has been found to be the case at north Assateague Island, Maryland, where sediment blockage by jetties has greatly increased the rate of landward barrier migration (Leatherman, 1984a), and the same qualitative result would be anticipated as a result of accelerated sea level rise.

Estuarine (Brackish) Marshes

Estuarine marshes embrace a wide variety of floristic species in diverse geologic settings where salinities are less than 30 ppt. These marshes, comprising integral components of major estuarine systems such as the Chesapeake Bay, occur in areas of quiescent waters and ample sediment supply. Accretionary budgets differ widely (Table 5-5), but in a dynamic equilibrium condition onsite production of organic materials and influx of mineral sediments cause vertical accretion, balancing the local rate of sea level rise. In view of the geographic range of the measurement sites and the local variability within coastal marshes, it is rather remarkable that the measured sedimentation rates all fall within the same order of magnitude. Accretion rates are generally found to vary from 1 mm/yr in high marsh at Duplin River, Georgia to 11 mm/yr in the Savannah River estuary, Georgia.

The data in Table 5-5 demonstrate that, in general, the measured accretion rates do exceed the locally determined relative rates of sea level rise. Consequently, most marshes do receive adequate sediment supply to compensate for current sea level rise. This must have been the case over the last few hundred years since "natural" marsh loss has not historically been reported to be a problem. Three notable exceptions occur at Barn Island, Connecticut; Blackwater Marsh, Maryland; and in Louisiana, where the present short-term rates of marsh accretion are lower than the local rates of sea level rise. In Louisiana there is widespread loss of coastal wetlands, in part attributable to a sediment deficit. Marsh deterioration is also known in the Blackwater Wildlife Refuge, Maryland, but no such problems are yet reported at Barn Island. Exceptionally low local rates of sediment accretion appear to be the cause in both Connecticut and Maryland.

Tidal Freshwater Marshes

Tidal freshwater marshes are located in the upper reaches of estuaries and other areas where ambient salinities are less than 5 ppt. The flora of these marshes is varied and lacks the typical vegetation zonation of open-coast marshes. The effects of rising sea levels will be saltwater intrusion and the eventual dominance of higher salt-tolerant plants. However, the effects of canalization on tidal freshwater marshes in the Mississippi delta demonstrate that dramatic increases in salinity over a comparatively short period exceed the capability of these marshes to adjust so that rapid losses ensue.

TABLE 5-5 Measured Marsh Accretion Rates (mm/yr) and Local Relative Sea Level Rise (mm/yr) along the Atlantic and Gulf of Mexico

Region	Salinity o/oo	Technique[a]	Accretion Range	Accretion Mean	Relative Rate of Sea Level Rise	Mean Tidal Range (m)
New England						
Barnstable, Mass.	20-30	H	3-8	5.5	0.9	2.9
Prudence Island, R.I.	28-32	L	2.8-5.8	4.3	1.9	1.1
Farm River, Conn.	--	L	--	5.0	1.9	1.8
Barn Island, Conn.	--	T		2.0	2.6	0.8
Great Island, Conn.	--	T		3.8	2.6	1.0
Hammock River, W., Conn.	--	T		3.6	2.6	1.4
Nells Island, Conn.	--	T		6.0	2.6	1.7
Stony Creek, Conn.	--	T		6.6	2.6	1.7
E. Hammock River, Conn.	(phragmites marsh)	T		17.1	2.6	1.4
Fresh Pond, N.Y.	26	L		4.3	2.2	2.0
Flax Pond, N.Y.	26	L	4.7-6.3	5.5	2.2	2.0
Mid-Atlantic						
Great Marsh, Del.	25-30	L		4.7	2.0	1.3
Lewes, Del.	25-30	L		> 9.4	2.0	1.3
Lewes Creek, Del.				5.0	2.0	1.3
Delaware (overall)[b]		C		1.45	~1.45	
Nanticoke, Md.	2-6	L	6.5-7.8	7.2	3.2	0.7
Blackwater, Md.	1-5	L	1.7-3.6	2.6	3.9	0.3
South-East						
North River, N.C.	--	L	2-4	3.0	1.9	0.9
North Inlet, S.C.	--	L, Cs	1.45-9.5	--	2.4	1.5
Savannah River, Ga.	--	L		11.0	2.5	2.3
Sapelo Island, Ga.	--	Cs	3-5	4.0	2.5	2.1

TABLE 5-5 (Continued)

Region	Salinity o/oo	Technique[a]	Accretion Range	Mean	Relative Rate of Sea Level Rise	Mean Tidal Range (m)
Duplin River, Ga.	--	T		9.0	2.5	2.5
Duplin River, Ga.	--	T		4.5	2.5	2.5
Duplin River, Ga.	--	T		1.0	2.5	2.5
Gulf of Mexico						
Barataria Bay, La.	< 1	Cs	6.4-10.6	6.9	9.5	0.5 (fresh)
Barataria Bay, La.	5-10	Cs	6.4-13.5	7.1	9.5	0.5 (brackish)
Barataria Bay, La.	10-15	Cs	5.9-14.0	6.7	9.5	0.5 (intermediate)
Barataria Bay, La.	> 15	Cs	7.5-13.5	8.1	9.5	0.5 (saline)
Calcasieu, La.	15	Cs	6.7-10.2	7.8	9.5	0.6

[a]Abbreviations: H: historical record; L: Lead ^{210}; Cs: Cesium^{137}; C: Carbon^{14}; and T: tracer and marker horizon.
[b]Long-term rates, averaged from many stations in Delaware for the entire available Holocene record.

SOURCE: Adapted from Stevenson et al., 1986

Processes of Marsh Loss with Sea Level Rise

Land losses in most marshes result from a combination of mechanisms. Shoreline erosion at the seaward edge of the marsh, being the most obvious process, could be expected to accelerate with increased water levels. Nationally, however, shoreline erosion probably accounts for about 1 percent of all marsh losses annually. The comparative resistance of marshy shorelines to wave attack suggests that with rapidly rising sea levels, most marshes will be long since submerged before extensive shoreline erosion occurs.

A more probable catastrophic mechanism of marsh loss with a large increase in sea levels will be the formation of extensive interior ponds allied with general tidal creek bank erosion and headward growth as tidal prisms increase. The rapid enlargement and coalescence of interior ponds in marshes subject to rapid coastal submergence has been amply documented in the Mississippi delta (DeLaune et al., 1983) and at the Blackwater Wildlife Refuge (Stevenson et al., 1986). The magnitude of marsh losses from interior ponding is instructive. At the Blackwater Wildlife Refuge in Maryland, over one-third of the total marsh area (about 5,000 acres) was lost between 1938 and 1979 by the growth of interior ponds, largely occurring during a 20-year period. The physiological mechanism behind the development of interior ponds is believed to be anoxia, and ultimate root death of marsh plants, as sea levels outpace the ability of the marsh to maintain elevation.

Human-Induced Changes

The most dramatic changes in wetlands have historically resulted from human alterations. Over half the salt marshes in New England have been lost because of dredge and fill activities. Elsewhere, the expanse of marshes has actually been increased by poor land practices. Early settlers felled large tracts of forest for agricultural fields, resulting in massive siltation of some bays and estuaries. This pattern is especially true of the Chesapeake Bay, where the colonial port at Gunpowder River is now separated from navigable waters by several miles of intertidal flats, colonized by marsh grasses.

Present human activities are mainly preventing sediments from reaching wetlands areas. Moreover, soil conservation practices through contour plowing, buffer strips, and no-till agriculture have substantially reduced the influx of soil into adjacent water

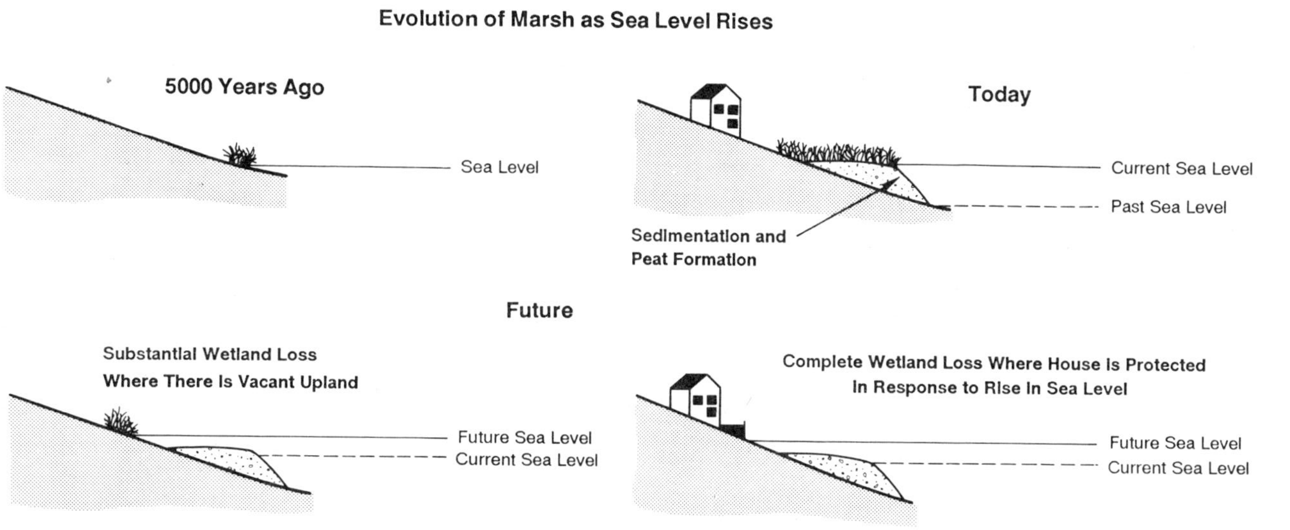

FIGURE 5-4 Coastal marshes have kept pace with the slow rate of sea level rise that has characterized the last several thousand years. Thus, the area of marsh has expanded over time as new lands have been inundated. If in the future sea level rises faster than the ability of the marsh to keep pace, the marsh area will contract. Construction of bulkheads to protect economic development may prevent new marsh from forming and result in a total loss of marsh in some areas. Source: Titus (1986).

bodies and wetlands. Dams and levees on major rivers trap material upstream and prevent over-bank flow of muds and fine sand during flood conditions. This is a particularly acute problem in coastal Louisiana, where the marshes have been established on deltaic sediments and have continued to accrete upward by sediments delivered during flood stage of the Mississippi River. Without these levees and other engineered structures, some cities built on floodplains, such as New Orleans, would be subject to massive and frequent flooding.

Finally, wetlands are being lost in coastal Louisiana because of pipeline and navigation canals that now lace much of the area, allowing saltwater intrusion. The resulting interactions are complicated, but there is no doubt that this practice has significantly contributed to the dramatic loss of wetlands presently being experienced in Louisiana.

Although salt marshes are protected by federal legislation, major losses of estuarine marshes can be anticipated in the future because of bulkheading along bay shores (Figure 5-4). With Holocene sea level rise, these salt marshes have been naturally translated landward through time. With the construction of landward-flanking bulkheads, which are prevalent along the mainland bay shores of many coastal states, these marshes will literally be squeezed out of existence with a sea level rise (Figure 5-4).

Prospects for Wetlands

The prospect for wetlands is bleak in light of existing conditions and projected changes. The present situation in coastal Louisiana can be used to forecast qualitatively the expected wetlands changes elsewhere. Due predominantly to subsidence from loading by the Mississippi delta and elimination of sediment supply by levee construction, the land surface has been subsiding about 1 cm/yr. Without the influx of massive quantities of inorganic riverborne sediments, the marsh surface can only accrete vertically by biogenic production, which is limited. Therefore, marshes are not able to keep pace with relative sea level rise (over 90 percent due to subsidence at present) and are being drowned in place. A rapidly subsiding substrate or accelerating sea level rise can yield similar results. Marsh grasses cannot accrete vertically fast enough to keep pace with sea level rise. This will likely be the fate for extensive estuarine marshes elsewhere in the United States if substantially higher rates of sea level rise are realized in the future.

6
Alternative Responses

As previously outlined, there has not been a long history of coping with sea level rise in the United States. Because of shore erosion, a portion of which is due to sea level rise, buildings have been lost and significant engineering projects have been undertaken during the past few centuries. Other countries have coped with relative sea level rise for thousands of years. Alternative responses to sea level rise, derived from worldwide experiences are described in this chapter.

COASTAL STRUCTURES AND PROTECTIVE TECHNIQUES

The performance and effectiveness of different types of coastal structures and protective techniques will be affected to varying degrees by a relative rise in sea level. Common to each type of erosion control structure under the action of sea level rise is the diminished efficiency due to submergence and overtopping. Structural failure becomes more likely as well, because wave forces can be greater due to the greater wave heights possible in deeper water and the higher-moment arm for the forces, providing greater fluid power.

Groins

These shore-perpendicular structures serve to reduce the local littoral drift rate, fostering sand impoundment on their updrift sides until they are filled to capacity, after which the longshore drift is allowed to bypass. If groins are allowed to fill from a natural sediment supply rather than from an alternative source as part of their construction, erosion of the adjacent shoreline will always occur. Groins are most effective along coastlines where a significant littoral drift occurs. They are often used to protect a long segment of coastline by the emplacement of a groin field.

The variety of groins in use, with differing lengths, widths, heights, permeabilities to sand, orientation, and spacing (between groins), has resulted in varying degrees of success in reducing erosion problems along the protected beaches. Examples of successful groin fields can be seen in such places as Rehoboth Beach, Delaware; Westhampton Beach, Long Island, New York; and Madeira Beach, Florida. However, Ocean City, Maryland has shifted away from the use of groins until a more complete understanding of all effects are known.

The beach downdrift of a groin field is often a location of accelerated erosion, and special treatment is necessary to protect this region. Often groin fields terminate at inlets, requiring no special measures; however, some groin fields terminate abruptly, requiring the use of beach nourishment, discussed below, or revetments of some kind. The erosion downdrift of the Westhampton Beach, New York groins shows the consequence of neglecting to provide for this effect.

The landward end of a groin typically extends into the dune line. As the sea level rises, the retreat of the dune line may leave the groin susceptible to flanking during high or storm tides, thus permitting sand to bypass the groin, reducing its effectiveness. The more readily the structure is flanked during normal weather conditions, the less the groin's sand-trapping and stabilizing capacity. Submergence of the groin by sea level rise brings on the same flanking effect, as well as overtopping of the groin by the longshore current and waves that transport the sand, again resulting in a loss of efficacy.

Groins constructed of durable material, such as stone, and appropriately designed can have a useful life exceeding 50 years. Using the three scenarios of sea level rise, the design of a rubble structure should include the capability to raise the crest elevation in keeping with the relative sea level rise. Groins constructed with wood, gabions, steel-sheet piling, and other less durable materials probably will have a useful life of less than 50 years. Therefore, no unusual measures accounting for sea level rise are needed at the present time in the design of these groins, using the adopted scenarios.

Bulkheads and Sea Walls

These structures are often used on shorelines above the mean high-water line to provide protection for the upland. Another use is to reduce flooding due to storm surges (e.g., the Galveston sea wall). These structures are often constructed as a vertical wall, facing the sea, thus occupying the least amount of land. A successful sea wall or bulkhead must be able to withstand not only the forces of incoming waves during a storm, but also the effects of overtopping, which permits a significant amount of water to add to the passive earth load exerted on the wall and can further result in a scouring or eroding of the backfill.

A common result of sea wall and bulkhead placement along the open coastline is the loss of the beach fronting the structure. This phenomenon, however, is not well understood. It appears that during a storm the volume of sand eroded at the base of a sea wall is nearly equivalent to the volume of upland erosion prevented by the sea wall. Thus, the offshore profile has a certain "demand" for sand and this is "satisfied" by erosion of the upland on a natural beach or as close as possible to the natural area of erosion on an armored shoreline. The practice of placing rubble at the toe of a wall to dissipate wave energy reduces or distributes this erosive effect.

As the mean shoreline retreats toward a bulkhead or sea wall as a result of rising relative sea level, the erosion in front of the wall is enhanced and overtopping increases. Dean and Maurmeyer (1983) provide a means, based on the concept of an equilibrium beach profile, to predict the amount of change in the beach profile due to changes in mean water level.

Sea level rise can be incorporated into the design of a sea wall in two ways. The first is to build the wall initially to account for the anticipated sea level rise during the life of the structure. Provided the freeboard is sufficient for the design life of the structure and it is engineered correctly for the forces it will experience, the sea wall should be immune to sea level rise effects. The other method is to design the wall with lower initial elevations (with less cost), increasing the elevation in the future as dictated by the relative sea level rise actually experienced and/or projected over relatively short (about a decade) time frames.

Revetments

A revetment consists of either loose or interlocking units laid on a slope, from the upland to some point on the profile, often below the depth of anticipated scour or fixed by a toe wall to prevent undermining by scouring. This structure serves the same objective as a bulkhead or sea wall, protecting the upland. While a revetment occupies a larger land area, the existence of a slope and the roughness provided by the structural elements may reduce the amount of erosion immediately seaward of the structure. Sea level rise and appropriate methods of accommodation are the same for revetments as for sea walls and bulkheads.

Beach Nourishment

Replenishing an eroding beach with sand is an effective means to restore a beach temporarily. Depending on the type and volume of nourishment sand, the temporary restoration may last for years. The massive (10.5-mile) effort along Miami Beach has lasted since 1980 without substantial volumetric erosion.

An attractive advantage of beach nourishment is that it is a soft solution to the erosion problem, i.e., no rigid structures are required. The drawback of beach nourishment, however, is that the processes that created the original erosion problem remain and continue to remove the nourishment sand. The length of time beach nourishment can be expected to last will depend on wave conditions.

Other factors that can influence the duration of a fill are the characteristics of the fill sand and the methods of placement. Sand that is finer than the original beach sand (particularly if it contains a significant silt fraction) will be eroded faster than the original sand. Fill not uniformly placed over the beach profile creates an out-of-equilibrium profile, which usually fosters offshore sediment transport, with attendant beach recession. Although this process of "profile equilibration" is accompanied by a shoreline recession and may be interpreted as an indication of poor performance of the project, in reality it should be viewed as an adjustment toward the natural profile with the recognition that the relocation sand is not lost, but remains in the nearshore system.

Using present technology, beach fill on stabilized shorelines will become more costly as sea level rises. As the offshore region deepens, the beach profile must steepen due to the fixed shoreline position. Using fill sand of the same grain sizes (or smaller) as the original beach sand will require far larger volumes of sand as the water level rises and the beach will become increasingly unstable. An alternative is to utilize coarser sand in future beach fills. Coarse sand permits a steeper beach profile and less transport offshore (Bascom, 1951).

A very approximate measure of the increased rate of losses can be developed by considering that the transport of sand away from the nourishment site is proportional to the wave height to the 2.5 power (Dean, 1976). The resulting percentage increase in beach nourishment volumes due to a sea level rise is

$$\left[\frac{(1+F)^{2.5}}{(1+F')^{2.5}} - 1\right] \times 100\% = 7\% \text{ (Case A)}$$

and

$$\left[\left(\frac{1+\Delta H}{H}\right)^{2.5} - 1\right] \times 100\% = 200\% \text{ (Case B)},$$

accounting for the effects of increased wave heights in the two examples presented in Chapter 4 (pp. 38–39).

It is of interest to examine the approximate costs of nourishment required to maintain the existing shoreline. This requires accurate projections of the rate of sea level rise. The calculations presented below will be based on two different formulations. First, Bruun's rule will be used for various sea level rise rates, which requires quantification of W (the active profile width) and h_* (the associated vertical dimension of this profile, including the berm elevation). Secondly, based on present (S_1) and projected (S_2) sea level rise rates, the ratio

$$\frac{R_2}{R_1} = \frac{S_2}{S_1}$$

can be formed, where R_1 and R_2 are the present and anticipated recession rates, respectively.

Method I

For illustrative purposes, consider the case of Florida's east coast. The long-term estimates of past relative sea level rise are 30 cm/century, and it is estimated that the limiting depth of motion h_B is on the order of 7 m and the berm height is 2 m, resulting in a h_* value of 9 m. The associated width could be determined from profiles or from the equilibrium beach profile

$$h_B = Ax_B^{2/3},$$

in which a representative value of A for this area has been determined from analysis of numerous profiles to be 0.1 $m^{1/3}$. Thus,

$$W = \left(\frac{h_B}{A}\right)^{3/2} = \left(\frac{9}{0.1}\right)^{3/2} = 854 \text{ m}.$$

Therefore, the recession rate multiplier for sea level rise, defined by the ratio of retreat R to sea level rise as determined from Bruun's rule, is

$$\frac{W}{h_*} = (584/9) = 95.$$

The present relative sea level rise rate is 30 cm/century, which appears to include a eustatic component of 12 cm and a neotectonic (subsidence) component of 18 cm. Assuming that the neotectonic component is unchanged over the next century, the relative rates of rise adopted in this report are as presented in the third column of Table 6-1.

The volume per unit length of beach V to maintain the shoreline position can be determined by considering a general form of the equation for the total retreat rate R_t, composed of retreat due to sea level rise R_s and advancement A due to additions of sand to the profile. The resulting equation is

$$R_t = R_s + A,$$

where the volume V required to result in an advancement $A = R_s$ such that the shoreline is stable, is

$$V = SW.$$

TABLE 6-1 Projected Shoreline Retreats and Costs of Maintaining the Shoreline by Nourishment over the Next Century on Florida's East Coast (Method I)

Scenario	Eustatic Rise (m)	Relative Rise (m)	Shoreline Retreat (m)	Average Annual Volumetric Requirements/ Unit Length (m^3/m)	Average Annual Costs[a]/ Unit Length of Shoreline ($/m)
I	0.5	0.7	45	4.1	33
II	1.0	1.2	78	7.0	56
III	1.5	1.7	111	10.0	80

[a]Costs are based on $8/$m^3$ and are 1987 approximate costs.

The annual volumetric requirements for the various scenarios are presented in column 5 of Table 6-1. These volumes are converted to annual costs using a 1987 cost of sand of $8/$m^3$. The ranges of annual maintenance nourishment costs associated with the three scenarios range from $33/m to $80/m of beach front. For comparison, the approximate range of values of beachfront property along the east coast of Florida is from $6,000/m to $60,000/m. The annual maintenance nourishment cost, expressed as a percentage of the value of the property, ranges from 0.06 to 1.3 percent. Thus, one could consider shoreline stabilization through nourishment as a "tax" or cost of living on a shoreline subject to natural erosive forces.

Method II

This method is much more direct. The annual average shoreline retreat rate R due to natural causes (relative sea level rise) along the east coast of Florida is approximately 0.5 m/yr. With the present sea level rise rate S_1 (= 30 cm/yr) and projected rates S_2, the projected retreat rates R_2 are

$$R_2 = R_1 \frac{S_2}{S_1},$$

and the associated required annual volumetric maintenance nourishment rates for shoreline stabilization V are

$$V_2 = R_2 h_*.$$

TABLE 6-2 Projected Shoreline Retreats and Costs of Maintaining the Shoreline by Nourishment over the Next Century on Florida's East Coast (Method II)

Scenario	Eustatic Rise (m)	Relative Rise (m)	Shoreline Retreat (m)	Average Annual Volumetric Requirements/ Unit Length (m^3/m)	Average Annual Costs[a]/ Unit Length of Shoreline ($/m)
I	0.5	0.7	117	10.5	84
II	1.0	1.2	200	18.0	144
III	1.5	1.7	283	25.5	204

[a]Costs are based on $8/$m^3$ and are approximate 1987 costs.

Adopting as before a value of $h_* = 9$ m, the results, including annual maintenance costs per unit length, are presented in Table 6-2. Summarizing briefly, the annual costs (in 1985 dollars) of stabilizing the shoreline range from $84/m to $204/m. For the same range of land values from $6,000/m to $60,000/m, this represents annual maintenance costs expressed as a percentage of value ranging from 0.1 to 3.4 percent. The shoreline stabilization costs through beach nourishment as predicted by the two methods differ by a factor of approximately 2.5 and reflect the inexact nature of this methodology.

Beach Nourishment with Groins

The use of groins with beach fill increases the time that the beach nourishment remains on the beach and reduces the downdrift erosion since the filled groins will begin to bypass sand immediately after construction. The response to sea level rise is the same as groins and beach fill, mentioned earlier.

Perched Beach

An interesting concept for rebuilding bathing beaches, the perched beach is an attempt to raise the local profile with fill and an offshore submerged sill that is oriented parallel to the shoreline. The sill is intended to retain the fill, simply acting as a "dam" or impediment to limit offshore sediment transport. The advantage of this technique is that beach fill is only required in the region shoreward of the sill, rather than along a large portion of the beach profile. The perched beach should be enclosed by shore-perpendicular structures, especially at the ends, to reduce the longshore loss of fill material. A test case for the perched beach has been carried out at Slaughter Beach, Delaware; however, no conclusions were drawn from the installation (U.S. Army Corps of Engineers, 1981).

The perched beach concept requires more testing because it consists of some design considerations that are poorly understood, such as the appropriate depth of water for the offshore sill. Also, it is not clear whether the offshore sills may act as a diode, permitting the loss of material in the offshore direction, but acting as a barrier to beach building by onshore transport of sand during favorable wave conditions.

Sea level rise will affect a perched beach in the same manner as beach nourishment, with the exception that the sill structure will become less efficient as the sea level rises, resulting in reduced sand retention. The sill should be anchored by shore-perpendicular return walls situated well inland in order to prevent flanking.

Offshore Breakwaters

The use of above-water, shore-parallel breakwaters to reduce wave heights at the shoreline and the potential for littoral drift is a very popular and effective international erosion control measure. In the United States, Winthrop Beach, Massachusetts; Lorraine, Ohio; and Presque Isle, Pennsylvania contain working examples of these structures, whose effectiveness is based on limiting the penetration of wave energy behind the breakwater. In Japan, more than 2,000 of these structures are in place (Toyoshima, 1982). Often a series of such structures is used; the spacing between breakwaters is an important parameter, as distance affects the amount of wave energy that passes to the protected beach.

Without shoreline stabilization provided by beach nourishment, rising water levels will effectively move the shoreline farther away from the breakwater, increasing the ability of the waves to diffract behind the structure and reducing the sheltering and efficacy of the device. Overtopping will obviously diminish the ability of offshore breakwaters to reduce the wave energy in the sheltered region. To be effective, designs must anticipate sea level rise, because the design lives of these structures are likely to be long. For example, they could be designed with higher initial top elevations or with features that make it possible to increase elevations in the future.

With increases in sea level, waves that attack the structures may increase in height, thus posing a greater threat. For example, the weight of stone, W_a, in a jetty or breakwater is chosen based on a design wave height. Using Hudson's formula in the *Shore Protection Manual* (U.S. Army Corps of Engineers, 1984), the design stone weight is proportional to the cube of the wave height. If the wave parameters in Chapter 4 are used, the increase in design stone weight due to relative sea level rise is

$$\frac{W_a'}{W_a} = \frac{(1+F)^3}{(1+F')^3} = 1.08 \text{ (Case A)}$$

or

$$\frac{W_a'}{W_a} = \left(\frac{1+\Delta H}{H}\right)^3 = 1.24 \text{ (Case B)}.$$

Thus, the increase in stone weight for these two examples would be 8–24 percent. The implication is that the margin of safety built into existing structures is reduced.

Storm Surge Barriers

Several barriers have been built in the United States to protect coastal cities from inundation during storm surges. Examples are the barriers at New Bedford, Massachusetts; Providence, Rhode Island; and Texas City, Texas. Others have been designed but not built (Perdikis, 1967). Internationally, probably the best known barriers are the Thames barrier, designed to protect the city of London, and the Delta Project to protect low-lying lands in the Netherlands. These barriers were designed with heights to exceed the surge elevations of certain design storms. As relative sea level rises, the factors of safety of these structures will be reduced.

Other Devices

There are numerous other devices used for beach erosion control. Several of them are available commercially but do not have the proven capability to eliminate or reduce beach erosion. Some of these devices are bottom mounted and would become more ineffective as sea level rises.

Polders are used in many countries for the reclamation of land from the sea. A polder, by definition, is land surrounded by dikes kept dry by the use of pumping. The Dutch have historically been the most active users of polders. In low-lying U.S. lands, as sea level rises and the need for land increases more use of polders may be made.

Effective management of estuarine sediments and sedimentation offers some potential for building up coastal wetlands. Dredged materials can be used to reinforce marshlands. Another alternative, especially in the Mississippi River delta, is to periodically divert sediment-laden river waters (usually contained behind levees) into marshes to allow natural deposition to take place.

Engineering Case Studies

The practicality of effective engineering response to increased future sea level rise can be addressed, in part, through the examination of case studies. Some facilities have been in place long enough to have experienced significant sea level rises. The case studies presented here include the Galveston sea wall and landfill at .ton, Texas; the Delta Project (dikes and surge barrier) in the Netherlands; the Harrison County, Mississippi beach nourishment project; Miami Beach, Florida beach nourishment; and the Tybee Island, Georgia sea walls, groins, and jetties.

Galveston, Texas

The city of Galveston is located on Galveston Island, a long barrier bounded on the east by the Bolivar Roads Inlet to Galveston Bay. In the late 1800s, Galveston was a summer resort community with extensive development. Existing sand dunes were removed for fill and beach access (Davis, 1961). The elevation of much of the island was extremely low; the average elevation in 1900 was 5.8 ft above mean low water (Engineering News, 1902).

On September 8, 1900 Galveston was demolished by a major hurricane. More than 6,000 people were killed and most of the buildings were flattened. To protect the city, a concrete sea wall, 16 ft high (with a crest elevation at 17 ft above mean low water), was constructed between October 1902 and July 1904. The wall characteristics included a curved face towards the sea and rubble toe protection to help dissipate wave energy and reduce wave scour. The sea wall was constructed on the beach along the +3 ft mean low-water contour (actually 1.6 ft above mean sea level). At this time, beach widths in front of the wall were up to 300 ft in some locations. The total cost of the wall was approximately $1.6 million, far less than the estimated $25 million in damage from the 1900 storm (Davis, undated).

In conjunction with the wall construction, the general elevation of the city was raised as an integral part of the plan to reduce storm surge flooding. Twelve million cubic yards of fill were placed by hydraulic dredging from a canal dug within the city limits and the bay, at a cost of $2 million (Engineering News, 1915). More than 2,000 homes were required to be raised by this plan. Additional fill was placed in 1909, after hurricane-induced wave overtopping removed some of the fill.

In 1915, another hurricane occurred with a storm surge of similar magnitude as the storm of 1900; it caused only 12 deaths and the property damage was $20 million less than in the 1900 storm. However, most of the beach was removed to an offshore bar and the beaches have continued to narrow since.

Over the years, the sea wall has been extended at both ends, to a total length of 10 miles. Maintenance has been required on the scour protection behind the wall after most storms, and the rubble toe protection has required additions as a result of subsidence into the sand and damage from six major storms since 1919. Subsidence of the wall is becoming a problem because part of the wall is located over a soft clay stratum. In one place the wall has subsided 1.4 ft.

Thirteen groins were constructed between 1936 and 1939, both to provide a beach for recreation and to protect the toe of the sea wall from scour. Although the groins have trapped a small amount of sand locally, no major accumulation has occurred.

Since 1904, the sea level at Galveston has risen approximately 24 cm, based on Leatherman's (1984b) interpretation of tidal gauge analyses of Hicks et al. (1983). Using the Bruun rule and Leatherman's figures, most of the shoreline loss can be attributed to relative sea level rise.

Leatherman (1984b) indicates that with the EPA high and midrange sea level scenarios for the year 2075 and a 100-year storm, the sea wall would be topped and the city flooded. For the high scenario, the 50-year storm would also flood the city. Clearly, the measure of protection afforded the city in the early 1900s is decreasing with increasing sea level. Leatherman indicates that diking will be necessary in the future to maintain the existing urban city. Other plans, such as providing a surge barrier to Galveston Bay, may be infeasible due to the low-lying islands fronting the bay.

The Netherlands

The foremost international example of a people coping with high relative sea level are the Dutch. Buffeted by catastrophic storm surges every several decades with the loss of thousands of lives and faced with the continuing subsidence of the land as the underlying peats and clays are compacted due to dewatering, the Dutch have spent centuries fighting the encroaching sea. Millions of people live below sea level at the present time and half of the country would be submerged without dikes.

From early settlement through the ninth century, the Frisians and the occupants of southeastern Holland, faced with periodic surge inundation by storms sweeping across the North Sea, built *terpen* and *vliedbergen*, both of which served the same purpose. *Terpen* were large areas of landfill on which homes and barns were built. Each mound contained about 1 million cubic yards of material and more than 1,260 were built (van Veen, 1962). *Vliedbergen* (hills of refuge) were large earthen mounds 10–12 m high on which people could wait out the floods.

After the ninth century, landowners began to band together to create dikes to protect existing upland from the encroaching sea. The frequent onslaught of storms provided the impetus to continue the diking effort. In one tragic example, 50,000 people lost their lives from a storm on December 14, 1287. In November 1421, 65 villages were submerged and 10,000 people were drowned (van Veen, 1962).

With the introduction of the windmill in the 1500s, serious reclamation was begun. Over 1,400 windmills were dedicated to pumping water out of low-lying areas. Not only were polders created to reclaim land lost to the sea, but also inland lakes, which were enlarging with the relative rise in sea level, were drained. In 1640, 27 lakes were drained under the leadership of Jan Leeghwater, a well-known Dutch engineer. From the thirteenth to the twentieth centuries, Holland reclaimed 1.3 million acres from the sea, but lost 1.4 million acres by the sea's encroachment. Without the reclamation efforts, the losses of land would have been much greater.

One of the major Dutch engineering works, begun in 1919,

was the reclamation of the Zuider Zee, which had been expanding constantly from its origin as a small freshwater lake (Lake Flevo) into a saltwater estuary. Although this project was controversial from the beginning due to its extremely high cost, the spirit of early engineers such as Andries Vierlingh, dike master to William the Silent, prevailed. Vierlingh wrote in 1570 in his treatise, *Tractaet van Dickagie*, as quoted by Wagret (1968), "The more one retreats, the more the sea prepares to expel one completely." The economic problems were difficult to overcome because the cost of the project exceeded the value of the reclaimed land. However, the benefits for future generations outweighed the merits of taking no action. In 1932, the Zuider Zee dike was completed and 550,000 acres of farm land were added to the Dutch nation (a 9 percent increase). Because of the high costs and environmental concerns associated with polders, not all of the sea bottom was reclaimed.

On February 1, 1953 the St. Ignatius flood, caused by a large winter storm moving across the North Sea, occurred with the loss of 1,850 lives and the flooding of many thousands of acres of crop land (almost 8 percent of the country) due to hundreds of breaches in the dikes, particularly south of Rotterdam. This massive storm created the pressure for the Delta Project, the world's largest coastal engineering work, which has resulted in the closing off of three major estuaries in the Rhine-Meuse delta region. This project will no longer permit intensive storm surge flooding.

The Delta Project consists of several phases. The first was the closure of the Haringvliet estuary, with the use of sluices to permit the efflux of Rhine River flows at low tide into the North Sea. The Grevelingenmeer was closed at both ends, creating a saline lake, with no apparent loss of water quality, to the surprise of most involved. Environmental concern about enclosed lakes led to the use of storm surge barrier gates for the largest of the estuaries, the Osterschelde. A total of 64 massive gates, which will be shut during major storms, permit tidal flows into the estuary to maintain existing water quality. The cost of the surge barriers (or *stormvloedkering*) is approximately $2 billion. A recent article in the *National Geographic* (October 1986) describes the construction of the barriers.

The design life of the Osterschelde barriers is 200 years, based on a design storm flood with a frequency of 1 in 4,000 years. This is a far longer design life and greater design storm than those used for any other coastal structure ever constructed.

The people of the Netherlands, with their limited land mass

and expanding population, have demonstrated that it is possible to defend against an encroaching sea, with its ever higher storm surges, using dikes and pumps. This has not been accompanied by a sacrifice of the beaches. Coastal resorts, located on diked islands, remain popular and are complete with bathing beaches. Examples are the beaches at Voorne, Goeree, Schouwen, and Walcheren islands in the Rhine estuary area. Additionally, the resort community of Scheveningen, nearly a part of The Hague now, has very wide beaches held in part by groins. In the north, the Frisian Islands beaches along the adjacent West German coast have been maintained in the face of relative sea level rise by migrating, as documented by Nummedal and Penland (1981).

Miami Beach, Florida

Between 1976 and 1980, a large beach nourishment and flood protection project was constructed by the U.S. Army Corps of Engineers at Miami Beach and cost $64 million. Over 14 million yd^3 of sand were placed on 10.5 miles of beach, from Bakers Harbor at the north end to Government Cut Entrance at the south. The resulting nourished beach averaged 300 ft wider than before. In addition to performing as a recreational beach, the project provides a flood and storm buffer for expensive property and rejuvenates the beach, the premier attraction of the city.

The fill material, dredged from offshore, had a large portion of fine and carbonate sands, leading to concerns about the stability of the fill. Measurements based on aerial photographs show that the shoreline at the north end of the fill retreated 100 ft within the first 5 years and remained stable over the next 4 years (up to 1985). It is likely that the initial shoreline retreat was a readjustment of the fill profile to an equilibrium profile.

The nourishment project has withstood some moderate hurricane activity (e.g., Hurricane David, 1979), and it has clearly met the needs of the coastal cities located behind the fill (Bal Harbour, Surfside, and Miami Beach).

Harrison County, Mississippi

The longest and one of the earliest beach restoration projects constructed was in Harrison County, Mississippi. This cooperative project, conducted by Harrison County with federal aid, encompassed some 26 miles of Mississippi Sound shoreline between Biloxi and Henderson Point. This area is sheltered by barrier islands from the direct attack of waves from the Gulf of Mexico.

The original project was constructed during 1951–1952 and included the placement of nearly 6 million yd^3 of fill from a borrow trench dredged to a depth of 15 ft and located about 1,500 ft offshore. The cost of the material placed was \$0.22 /yd^3, and the project resulted in some 700 acres of new beach with a width in excess of 300 ft and a berm height of 5 ft. A sea wall some 25 miles long had been constructed during the years 1925–1928 to protect property and highway U.S. 90, immediately upland of the sea wall. The longshore transport along this beach is from east to west. A terminal structure is located at Henderson Point at the entrance to Bay St. Louis, the western (downdrift) end of the project. Numerous concrete drainage trenches were constructed across the beach and function as groins, thereby helping to stabilize the placed beach.

This project is generally considered to have performed well. Annual losses were estimated to be on the order of 100,000 yd^3/yr, with a considerable portion of this amount due to sand being blown inland.

In 1969, Hurricane Camille, one of the two most intense storms on record in the Gulf of Mexico, made landfall near the western end of the project, causing record storm tides in excess of 22 ft and, understandably, causing some sand losses.

During 1972–1973 the project was renourished with 1.9 million yd^3 of sand. The project was inspected in the summer of 1985, prior to Hurricane Elena, and appeared to be performing well. Undoubtedly, this is an example of a project that has provided both protection to the upland against severe storms and a valuable recreational facility. Based on the data of Hicks et al. (1983) the estimated relative sea level rise over the period encompassing the beach restoration project (1952–1985) is approximately 8 cm, too small to be indicative of the stability of a nourishment project in an era of sea level rise and in the presence of substantial longshore sediment transport. In comparing the relative longevity of this project with others, one must consider the sheltering provided by the offshore islands that form the gulfward boundary of Mississippi Sound.

Tybee Island, Georgia

Tybee Island, Georgia is a barrier island some 6 km long located just south (downdrift) of the entrance to the Savannah River.

Navigational improvements to Savannah River include jetties and a deepened channel that have effectively eliminated any sediment supply from the north. This lack of sediment supply is reflected, in part, by the landward migration of offshore contours and the erosive stress on Tybee Island.

Tybee Island represents an interesting case study due to the long history of erosion studies and variety of erosion control measures employed. The earliest studies date back to 1855. Erosion control measures have included shore parallel structures (revetments and sea walls), groins, and beach nourishment. Shoreline positions documented by these studies are presented in Figure 6-1.

In 1882, three rock groins were constructed at the north end of the island, although it is not evident whether these were for erosion control or river training. Between 1912 and 1930, several additional groins and portions of a sea wall were constructed.

In 1931, additional erosion control efforts were initiated, including a 2,650 ft long bulkhead and 5 groins extending from the bulkhead. Numerous structures were tried, and in the late 1930s and early 1940s a concrete sea wall was constructed extending along the entire length of Tybee Island. Hurricane Dora in 1964 caused failure of a portion of the sea wall. This failed section, approximately 1.5 km in length, was protected by a rock revetment.

A Corps of Engineers study culminated in 1971 with the recommendation for three substantial groins and a beach nourishment project. The sand was to be placed at the north end of the island, with one groin to be located at the northerly limit of sand placement and the other two near the north end and center of the project. This project was constructed in the period 1974–1976. However, only the northerly structure was built; it extended 800 ft from the sea wall. Total sand placed was 2.26 million yd^3. The borrow area for the project was a shoal extending southeast from the island. The project performance was monitored and initial results indicated more rapid losses than anticipated. These early losses from the project areas occurred (1) over and through the permeable north groin resulting in 10–12 acres accumulation of dry sandy area on the north end of the island, and (2) at the south end of the island where material appeared to be "drawn" to the substantial depression resulting from the borrow operation. The center of the island accreted.

In summary, the erosion stresses at Tybee Island are abnormally high due to the navigational works at the entrance to the Savannah River. With more than 100 years of erosion control

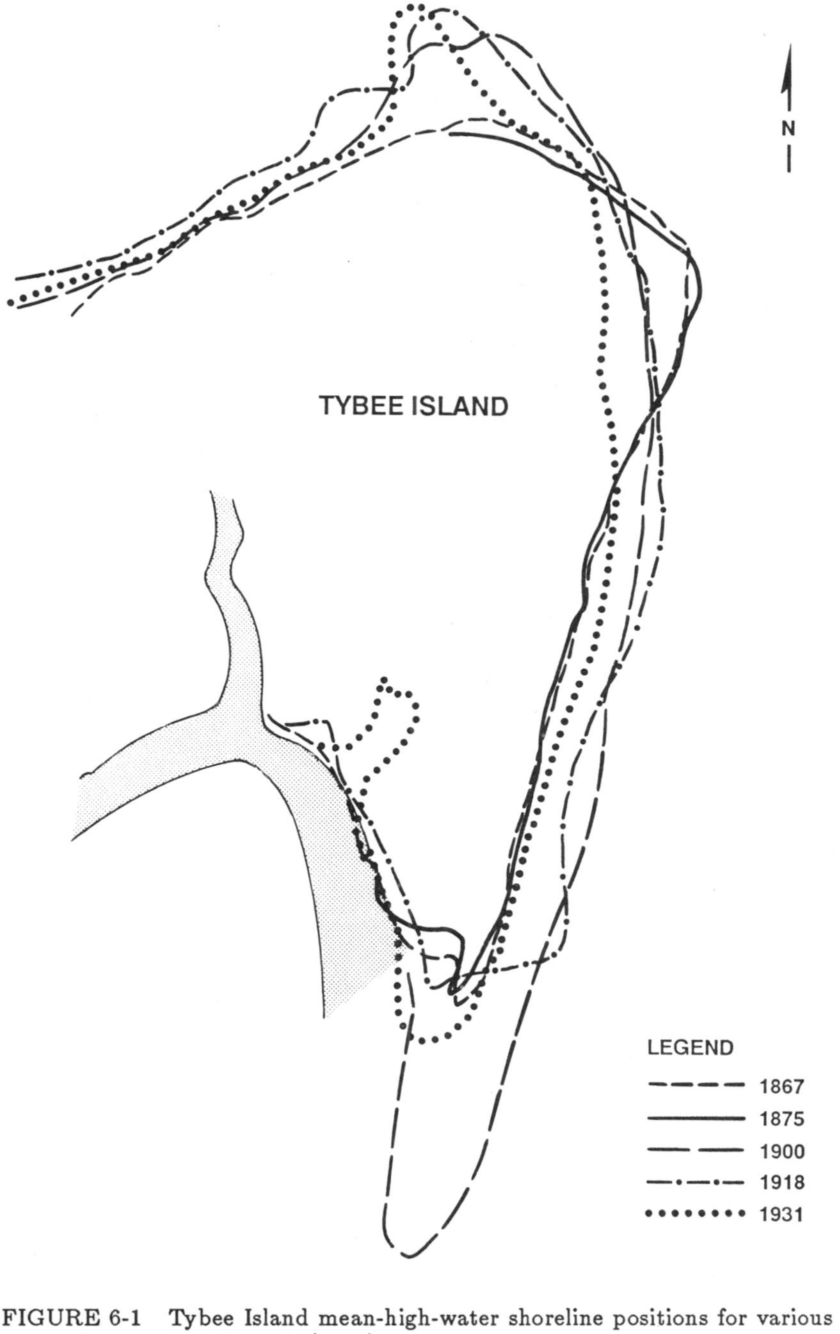

FIGURE 6-1 Tybee Island mean-high-water shoreline positions for various years. Source: Oertel et al. (1985).

efforts, during which relative sea level has risen over 40 cm, the shoreline has not eroded as much as might be expected and the erosion control efforts have been moderately successful to date.

The historic shorelines (Figure 6-1) have experienced substantial fluctuations, but the dominant changes between 1867 and 1931 were (1) the loss of a projection near the northeast end of the island, and (2) the deposition near the north of the island. The areal changes in these two features appear to balance approximately. The sea wall construction program completed in the early 1940s "fixed" the shoreline position against severe storms. Beach nourishment during 1975–1976 contributed to the formation of recreational beach areas, still present after 10 years. Near the central portion of the island, sand has accumulated, resulting in a fairly substantial dune field up to 70 m wide.

Terminal Island, California

The extreme rise in relative mean sea level experienced at Terminal Island and a portion of Long Beach, California some years ago was dominantly due to subsidence (Allen and Mayrega, 1970). The first evidence of the phenomenon occurred in the late 1930s and early 1940s when surveyors began to have difficulty in reproducing leveling measurements. The discrepancies became so prevalent that the U.S. Coast and Geodetic Survey was called upon to run a new first order survey from the mountains on either side of the Los Angeles basin across the waterfront. The results showed that in the few years since the last set of levels, an area about 3 miles wide and 4 miles long had subsided about 4 ft, in a dish-shaped depression. The center of the depression was near the eastern end of Terminal Island, where the largest steam electric-generating plant in southern California was located.

A series of studies was commissioned that conclusively identified the major cause of subsidence as the withdrawal of oil and gas from the Wilmington Oil Field, the limits of which closely matched the subsidence contours. The recommended remedy was to maintain pressure in the various strata comprising the field. This was accomplished by organizing the oil field so that some of the existing wells could be used for production and others for water injection to maintain pressure, and such that the water would "sweep" the oil to the production wells. The hydrocarbons produced were shared by all companies with a financial interest in the field.

By the 1970s subsidence had been arrested and a small rebound had even occurred. The overall subsidence ranged up to 20 ft with considerable damage to harbor facilities, pipelines, culverts, buried cables, and other structures. This damage required substantial remedial efforts, including diking in areas of extreme subsidence, reconstruction of damaged facilities, bridge repair, and redrilling of oil wells that had experienced casing damage.

The experiences in the Long Beach/Terminal Island area resulted in measures to counter an extreme relative mean sea level rise. The changes occurred much more rapidly than those expected with rising relative sea levels elsewhere, and they included horizontal movements of points on land that would not be expected with a general rise in sea level. This experience illustrates the nature and effectiveness of some of the measures that may be needed along the sea coasts.

RETREAT

Holding back the sea as water levels rise will almost always be technically feasible; however, in some cases it may not be economically or environmentally sound. In areas where the long-range cost or environmental damage due to shoreline stabilization is unacceptable, it will be advisable for development to retreat or move back from the shore. Although stabilization measures can be deferred until an accelerated rise makes moves necessary, a planned decision to retreat would require a lead time of years.

A retreat can occur as either a gradual process or as a catastrophic abandonment. Examples of the former would include removing buildings as they are threatened or as they interfere with use of the beach, and avoiding major renovations of buildings or new construction that would soon be threatened by higher sea levels. The latter might involve prohibiting the reconstruction of buildings destroyed or damaged by storms. This approach is being taken on Galveston Island by the state of Texas in the wake of Hurricane Alicia in 1983.

A recent conference of coastal scientists, engineers, and policy analysts (Howard et al., 1985) concluded that it may be preferable for some communities to move back from the shoreline in a planned and orderly fashion. Otherwise, as sea level rises there is a significant likelihood that a number of communities will retreat involuntarily as a result of unpredictable disasters.

Mechanism of Retreat

There are three basic ways to retreat from an eroding shoreline: (1) buildings can be moved as the shoreline approaches, (2) buildings can be written off and the remnants removed after being destroyed in storms, or (3) the construction of buildings near beaches can be avoided altogether.

An example of the third approach is the anticipatory land-use planning for erosion in North Carolina. A movable house must be set back from the shore the distance of the erosion expected in the next 30 years; immovable buildings, such as high rises, must be set back a distance equal to 60 years of expected erosion (North Carolina Office of Coastal Management, 1984). In Maine, new buildings must be set back far enough to permit 100 years of erosion. Both states assume that current erosion trends will not accelerate as a result of projected sea level rise.

North Carolina and Maine have essentially chosen a policy of gradual retreat from the shore. Both states have enacted regulations prohibiting placement of hard structures of any kind on eroding open-ocean shorelines. In 1984, 27 erosion-threatened buildings were moved back from the North Carolina shore; the regulations will be put to a more severe test in the future, when multistory condominiums are threatened by erosion.

Putting a policy of retreat in place can be accomplished in various ways by different communities. Areas with low-density coastal development can rely on building codes, setbacks, zoning, and land-use plans. More developed communities will have to address the issues of existing buildings and shoreline stabilization structures. The problems are so diverse that their solutions will require many different actions by different levels of government as well as the private sector. The diversity of retreat mechanisms will be governed by the widely varying characteristics of natural shoreline systems.

Some of the methods government might use to prepare for retreat are included in Howard et al. (1985). Of those recommendations for implementing retreat, the ones most related to engineering issues follow:

1. Halt stabilization of the shoreline. No more funds should be used to hold the shoreline in place under the retreat alternative.

2. Establish construction setback lines in states that do not have them. Seaward of these setback lines, no construction can be permitted. Setback lines exist in Florida, Maine, North Carolina,

Alabama, and Delaware, to name a few states with the necessary enabling state legislation. Furthermore, for rapidly eroding shorelines, a time-dependent setback line may be established to allow for further retreat as shorelines recede.

3. Remove coastal stabilization devices that become threats to public safety, as well as structures, including buildings, that become undermined by the sea.

4. Encourage further work in coastal processes research to provide greater scientific backing for the design of setback lines, as well as to develop innovative technologies for sand bypassing at inlets and development of cost-effective coastal protection schemes.

Implicit in the philosophy of retreat is the belief that cost-effective coastal protection is not viable for the given locale. Since the state of the art of coastal erosion mitigation is evolving rapidly, any retreat decision should be reviewed periodically. If the benefits of shoreline stabilization exceed its costs, then the retreat decision should be reevaluated.

Engineering, Geologic, and Economic Considerations

A decision to retreat or not and the choice of retreat mechanisms should be based on a sound understanding of coastal processes. Perhaps the single most important such consideration is the impact the actions of one community can have on neighboring communities whose beaches are connected to the same sand supply system. To reduce the potential for sand loss and damage to recreational beaches, communities that do not choose to retreat should ideally be at the terminus of the sand supply line for a given coastal reach. For example, stabilization of eroding bluffs or headlands should be discouraged if it can be demonstrated that beaches in adjacent communities will suffer as a result of the loss of eroding material. In general, sources of sand should not be stabilized; areas near sand sinks are much more suitable for stabilization by devices such as sea walls and revetments. Recognizing that a retreating shoreline provides a sand source to downdrift shorelines, in situations in which shoreline stabilization is deemed justified, the state of Florida requires annual mitigation through sand placement in the beach system to offset the material prevented from entering the system through natural erosion processes.

Clearly, if some segments of the shoreline remain in place and others are allowed to move back in response to a rising sea level, major shoreline offsets will eventually occur. The evolving nearshore system and its impact on sand supply must be a major consideration in the design of a retreat scenario.

Without political or emotional considerations, economics will be the final arbiter in deciding whether or not to retreat. Each governmental entity generates a certain amount of disposable income. If the cost of stabilizing the shoreline is a small percentage (say 5 percent) of the disposable income, then stabilizing may be appropriate. However, if stabilization requires the expenditure of a large percentage of the disposal income, then retreat may be the appropriate alternative.

CONTROLLING THE RATE OF SEA LEVEL RISE

In contrast to the defensive approach to sea level rise, a concept for control through active alteration of the hydrologic cycle by management also has been suggested (Newman and Fairbridge, 1986). The strategy involves storing water in unfilled pore spaces of sedimentary rocks in the earth's arid regions. According to Newman and Fairbridge, "The world's greatest surface water storage potential lies within the Eurasian land mass: the Aral-Caspian sink." Additional inland basins exist in other areas of the world. The concept also involves river diversion, and, potentially, subsequent hydrologic recycling methods.

It is noted that such water diversions would affect ocean chemistry and dynamics as well as ecological conditions on land. Neither the environmental consequences nor the practicality of implementing the strategy is addressed. Diversion of rivers that flow into the Arctic Ocean and Bering Sea has been discussed by Soviet Union engineers in past years to address a variety of regional problems. Uncertainties about the environmental consequences have been raised, and the proposals have not been developed for implementation (Sullivan, 1986).

Another approach is to limit the global warming due to the greenhouse effect. The United States banned the use of chlorofluorocarbon in spray cans in the 1970s, and international negotiations are under way to further limit emissions of this gas. Control of carbon monoxide emission would indirectly limit increases in methane because CO_2 depletes atmosphere OH^-, which removes methane from the atmosphere. A number of political leaders have suggested that CO_2 emissions be reduced through an end to global deforestation and a gradual shift from fossil fuels to solar and nuclear energy, which do not emit CO_2. Even a shift from coal to natural gas would decrease CO_2 emissions significantly. Nevertheless, the time that it would take to replace completely our fossil fuel infrastructure suggests that it will be very difficult to limit the global warming expected in the next several decades.

7
Assessment of Response Strategies for Specific Facilities and Systems

Many facilities and systems in coastal regions will be affected by changes in relative mean sea level. The effects and their importance will vary substantially depending on the type of facility and its location. Some structures on the exposed ocean coast are designed to prevent or decrease relatively long-term erosion, to protect buildings, roads, and other facilities during intervals of severe cyclic erosion, or to create wider beaches. These systems have been previously addressed. This chapter considers airports; levees and canals; seaports; port structures; navigation channels, turning basins, docking areas, and navigation gates; piers and wharfs; dry and wet docks; highways, railroads, vehicular tunnels, and bridges; storm drains, pipelines, and upstream water quality systems; flood control; commercial and industrial buildings; power plants and associated cooling water systems; hotels and malls; and residential centers.

In the design, construction, and utilization of many of these facilities, a "working and economic life" of 50 years or less is commonly considered. However, functional needs change, facilities wear out, and equipment becomes obsolete. Using the three scenarios for relative mean sea level rise discussed in Chapters 1 and 2 of this report results in a range of rise during the next 50 years of about 6 in. to slightly more than 1 ft. The engineering implications of this change would be moderate. For projects with much longer life expectancies, however, there are major engineering implications.

Some preliminary studies have examined the effects of a rise in relative mean sea level. For example, the Hawaii Coastal Zone Management Program (1985) studied the effects on Honolulu, Hawaii using four scenarios. For each scenario the report presented a map showing the projected shoreline/wetland line and the location of the present and projected coastal flood hazard zone. Possible effects on the port, the airport, Waikiki Beach, streets, and buildings were also addressed. The study concluded that owing to uncertainties relating to predicted eustatic mean sea level rise: (1) scientific predictive capabilities should be monitored, and (2) a meeting should be held in mid-1989 to reconsider the situation and make specific recommendations.

Some constructed facilities and systems might require the installation of levees (dikes) to protect them. Experience gained in protecting cities from the ocean in cases of relatively rapid subsidence is valuable. The case study of Long Beach/Terminal Island, California is a good example. More extensive experience has been gained in Japan, where subsidence was caused primarily by groundwater withdrawal. In Tokyo, for example, the cumulative subsidence up to 1968 in the eastern section of the city, facing Tokyo Bay, amounted to a maximum of 4.2 m. An extensive area became a lowland, with about 115 km^2 below sea level. The project reported by Tagami et al. (1970) includes 253 km of tide levees, 41 sluice gates, and 9 pumping stations. At the time the paper was written, 75 km of levees had been completed.

Another example is Osaka, the second largest city in Japan, situated on the alluvial plain of the Yodo and Yamato rivers. Between 1935 and 1968, the land in the vicinity of the Port of Osaka subsided by amounts ranging from about 50 cm inland to about 250 cm in the harbor area. To protect the city, 124 km of levees were constructed along the coast of the bay and on both sides of the rivers running through the city, with a crest elevation of 5 m above datum (the lowest low-water level observed in the port in 1885). Presently, there are plans to increase the elevation to 6.6 m (Murayama, 1970).

AIRPORTS

The airports of many coastal cities are constructed on landfill in bays (e.g., the airports of San Francisco and Oakland, California, La Guardia Field, New York, and Boston, Massachusetts). Some levees (dikes) used to protect the airports from the bay water are at minimum elevations and of minimum construction standards, so that a few feet increase in relative mean sea level would result in levee overtopping during severe storms, with the possible breaching of some sections. The degree of the eventual problem is specific to the site. Normal maintenance often requires placing more material on levees to compensate for settlement and consolidation of core materials and foundations. Needed adjustments resulting from relative mean sea level rise may be made as a part of this maintenance.

A study by the Hawaii Coastal Zone Management Program (1985) concluded that a 1.9-ft rise in relative mean sea level would result in temporary disruptions of transportation at Honolulu International Airport; a 4.8-ft rise would cause frequent and prolonged disruptions if no remedial works such as levees were emplaced.

Drainage problems exist in low-lying airports and will worsen with a rise in relative mean sea level. In addition, there are wetlands in some airport properties that require pumping and an increase in relative mean sea level would necessitate more pumping. However, in such cases, conflicts of interest may arise between airport authorities and other agencies, such as the U.S. Wildlife Service.

LEVEES

Existing Levees

Many miles of levees along bays and tidal rivers would be affected by a rise in relative mean sea level. One example of present hazards to levees is the Sacramento-San Joaquin delta, California (Figure 7-1). According to the California Department of Water Resources (1983b):

> Since 1980, levee failures have occurred on 12 of about 60 Delta islands. Factors that contribute to levee failures include: instability of the levee section and foundation materials; subsidence; rodent burrows; erosion from wind waves and boat wakes; inadequate

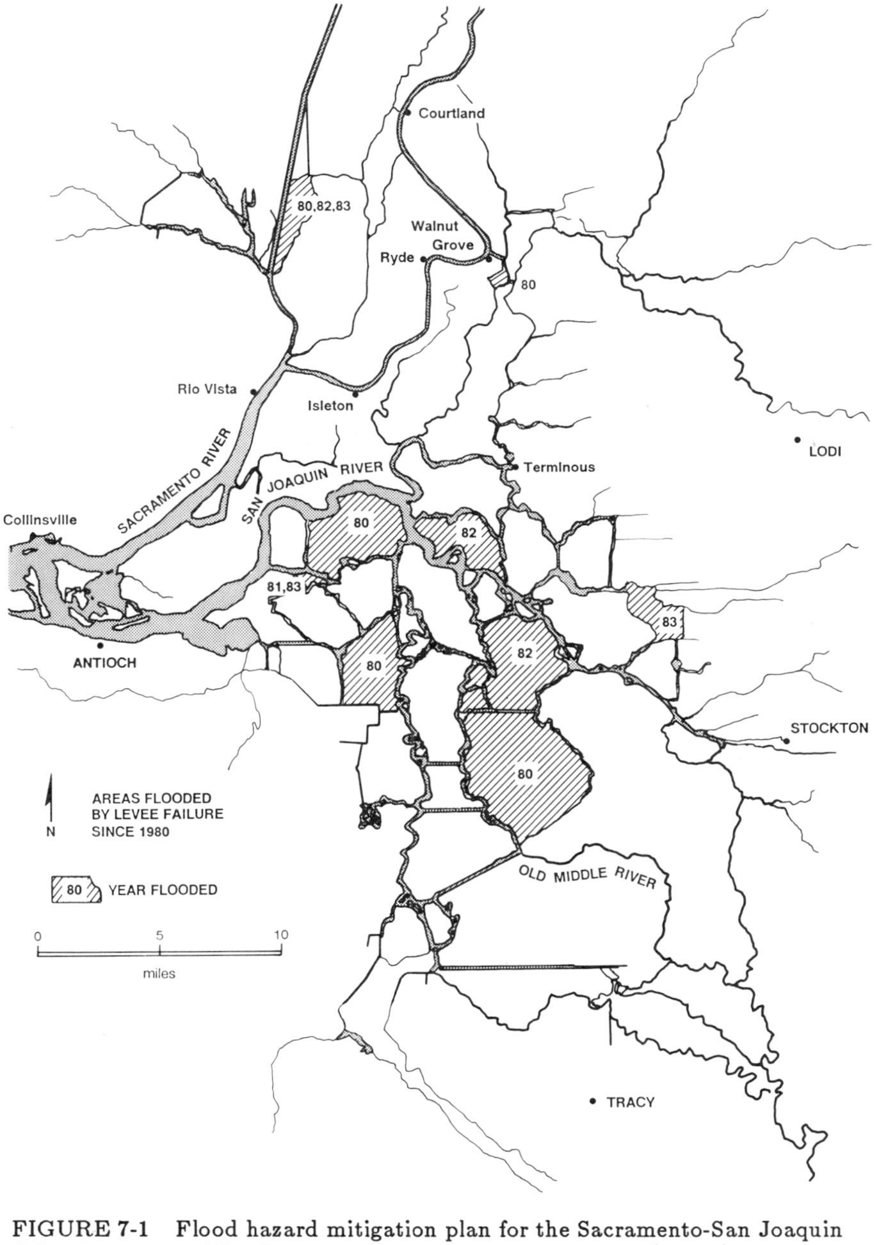

FIGURE 7-1 Flood hazard mitigation plan for the Sacramento-San Joaquin delta. Source: California Department of Water Resources (1983b).

height (freeboard); seismic activity; and seepage. Flooding of islands can have several adverse impacts, including temporary detriments to water quality due to ocean water intrusion, increased loss of water by evaporation, increased seepage on islands adjacent to the flooded areas, loss of agricultural land, damage to urban and recreational developments, and fish and wildlife losses.

A letter of February 13, 1986 from Mr. David N. Kennedy, director of the California Department of Water Resources, indicates that since 1980, about $100 million of emergency funds from federal, state, and local sources have been spent shoring up delta levees and reclaiming flooded islands, and that the state is now furnishing about $2 million each year to improve the levees.

It appears that a 1-ft increase in relative mean sea level could have a major impact on the protective capabilities of the levees. Increases in levee elevations (and base widths) should be made if and when needed and justified economically. The added weight of the levees on the soil would increase the rate of subsidence, and this would require design and construction consideration. Earthquake resistance should also be accommodated in seismically active areas.

Many of the levees were constructed years ago when it was relatively simple to obtain and transport material for their construction. Additional information may be needed to identify problems that could be encountered in the future and to develop potential solutions. Issues include sources and transport of material, environmental conflicts, and requirements for widening the base.

In some farming areas surrounded by levees, the soil type is such that the ground level becomes lower with time as a result of farming, wind erosion, fires, biochemical oxidation, and subsidence due to compaction (Stephens and Speir, 1970). These effects, combined with an increase in eustatic mean sea level, will change levee stability.

Preliminary information from the Delta Levee Subventions Program indicates that a long-term data collection program is desirable to measure rates of subsidence (Kennedy, 1986). Deep-measurement compaction recorders may be used to permit the separation of surface oxidation and compaction of the peat soil from deeper subsidence resulting from water and/or gas withdrawal. Use of satellite surveying methods is planned to determine ground elevations, since concern exists that much of the data obtained with standard leveling techniques in the past is invalid because of land instability in the delta.

The California Central Valley is quite flat, and the Sacramento River is tidal as far upstream as Sacramento. As a first approximation, the levees would have to be increased in elevation by the same amount (slightly more in bay areas where deeper water allows higher waves to be generated) as any increase in relative mean sea level.

Gates are an integral component of some levees in the United States; examples include the Fox Point Barrier in Narragansett Bay, the New Bedford Hurricane Barrier, the hurricane barrier system at New Orleans, Louisiana, and the Texas City Hurricane Protection System. Whether a 0.5- to 1.0-ft rise in relative mean sea level during the next 50 years would require modification of the gates is unknown. Study of the typhoon gates in Japan might provide information concerning this issue.

New Levees

The construction of new levees may be a solution to protect some densely populated areas from a substantial rise in relative mean sea level. The National Research Council (1983) considered the case of a very large rise (5 m) in relative mean sea level for Boston, Massachusetts. This rise is much greater than any of the values considered in this report, but it is noteworthy that a practical solution may be possible in some areas. The report states (pp. 473–474):

> A rudimentary illustration of the economics can be based on the Boston area. A full 5 m would cover most of downtown Boston. Beacon Hill, containing the State House, would be an island separated by about 3 km from the nearest mainland. Most of adjacent Cambridge would be awash. But it would take only 4 km of dikes, mostly built on land that is currently above sea level, to defend the entire area. Perhaps even more economical, because it would avoid the political costs of choosing what to save and what to give up and of condemning land for right-of-way, would be a dike 8 or 10 km in length to enclose all of Boston Harbor.
>
> If that were done, new deep-water port facilities would have to be constructed outside the enclosed harbor; locks would permit small boats in and out. The Charles and Mystic Rivers would have to be accommodated. Whether in a couple of hundred years there would be any significant flow in those rivers would depend on changing climate and increasing demand for water. Levees, a diversion canal, or pumping could be compared for costs, and ecological impacts.

The study cited above could be redone to determine if any levee would be needed or justified for a 0.5- to 1.0-ft rise during the next 50 years. The dike length of 8–10 km mentioned in the above study is very small compared with existing levees in Tokyo and Osaka, Japan.

Major problems will be associated with new levees, including land condemnation, sources and transportation of construction material, and environmental conflicts. In addition, navigation gates may be needed in some regions.

SEDIMENTATION OF SEAPORTS AND HARBORS, NAVIGATION CHANNELS, TURNING BASINS, AND DOCKING AREAS

In a Marine Board document (NRC, 1985a), it is stated that there are 102 ports in the United States serving oceangoing traffic (defined arbitrarily as 30 ft or more in depth). Twenty-six of these are primarily shoreline or coastal, 55 are in estuaries, and 21 are basically river ports (not entirely exclusive categories). No pattern of equivalence was found to exist among the ports and harbors. Each has its unique set of conditions of topography, bathymetry, tides, current and wind variations, temperature and climate, salinity and turbidity, and sediment transport regimes, which, with the man-made developments, result in a different situation for each case.

Changes in tidal and other currents will occur with changes in relative mean sea level, and changes in tide range and phase may occur. These may, in turn, cause changes in siltation. Hydrographic surveys will be required, and changes made in bathymetric charts, tide tables, and current tables.

The alteration of seiching conditions could be locally significant. Wherever the coastline tends to focus long waves of a certain period in the manner of a lens, a rise in sea level may produce disproportionately increased seiche heights. New resonant locations may develop, and old ones disappear. The effects of these changing conditions on moorings and cargo handling need to be studied.

One of the findings of the Marine Board (NRC, 1985a:5) is

> It is possible that major improvement in dredging will increase deposition rates in certain locations within a harbor rather than reducing the problem. This is because deepening certain channels, especially in estuaries allows seawater intrusion farther upstream

> than before dredging, which can cause deposition of fine sediments (which normally would be carried seaward) in the upper reaches.

This statement was made for the case of no sea level rise. What would be the effect of a rise in relative mean sea level? At first glance it might be thought that it would result in deeper navigation channels, turning basins, and docking areas than at present, which would result in a decreased need for dredging. This might be true for the short term, but not necessarily in the long term, as is implied by the above statement. Major changes would probably occur in the location of the saltwater wedge, and thus in the location of shoaling of the river channel. Sediment movement upstream would affect the costs of dredging and disposing of dredged material. Because of the complexity of siltation and the high costs of dredging, greater detail is devoted to this issue in this report.

Consider some of the effects on sedimentation of a rise of relative mean sea level (recall that there are 55 ports in estuarine environments in the United States). As the mean sea level rose after the last Ice Age, the sea invaded valleys along coasts created in the past by fluvial erosion and tectonic processes. Whenever a river draining a valley carried enough sediment to fill the drowning valley at the rate of sea level rise, a river delta developed (e.g., Mississippi River).

Very different estuary configurations evolved where there was insufficient sediment to fill the valley as sea level rose. These occurred where the valley was wide, or where the stream carried little sediment. The invading sea created large shallow bays that deepened as sea level rose, as in the San Francisco Bay system, Delaware Bay, and the Chesapeake Bay. Large, shallow bays provide hydraulic conditions that facilitate sedimentation of riverborne sediments and deposition of sediments brought to the bays during high-flow events. Sediments accumulate at the river deposition sites and gradually extend into the bay (Krone, 1979).

Daytime onshore breezes are typical of estuaries in the summer, and these breezes generate waves on the bays. The ability of waves to suspend sediment increases rapidly with decreasing water depth. Many estuaries are shallow enough that the wave action generated by onshore breezes can suspend deposited material and hold it in suspension while tidal currents circulate it throughout the system. At night, when onshore breezes die, the suspended sediment settles. If it settles where subsequent wave action or tidal currents resuspend it, it continues to circulate and may continue to the sea. Under these conditions the upper bays fill to the level where, over time, suspension by wave action equals the supply of riverborne sediment (Krone, 1979).

The supply of riverborne sediment has typically increased with land development. For example, sediment supply to the San Francisco Bay system was vastly increased from the 1850s to 1877 by extensive hydraulic mining in the Sierra Nevada foothills. About 1.9 billion yd^3 of material deposited in the upper bay areas of the San Francisco Bay system (Gilbert, 1917), which filled to the level at which wave action maintained the water depth. Sea level has risen about 0.7 ft since the hydraulic mining era. Sediment supply after this period apparently continues to be greater than that prior to 1850, and the upper bay areas are shallow, so that the deposition is limited by wave action. The extent of the deposit continues to progress toward the estuary mouth (Krone, 1979).

Another consequence of sea level rise in drowned valleys is the development of marshes along the shore (Krone, 1985). Examination of historical sea levels shows that the rise has not been continuous, but has fluctuated through time. Along the shore of a bay, tidal flats that developed during a period of higher sea level are exposed for a greater portion of the tidal cycle when sea level temporarily falls. Some types of plants become established on these mud flats when they are slightly above mean tide level. Plants trap suspended sediment when they are inundated by tides and reduce erosion by waves, so that the rate of sedimentation on the marsh is enhanced. The elevation of the marsh surface rises rapidly, gradually slowing as the increasing elevation reduces the frequency and duration of flooding by the tide. Depending on the rate of sea level rise and the supply of suspended sediment, the marsh surface tends to maintain its level relative to the tide as sea level rises.

The rising marsh surface caused the marsh to invade land where the shoreline slope is gradual, and over time extensive marshlands developed. Such marshes were significant traps for suspended sediment and undoubtedly affected the rates of deposition in the bays. In developed countries many estuary marshes have been diked for use as salt evaporation ponds or filled for agriculture and urban development, and thus no longer play a part in estuarine hydrology. Enough information to calculate the loss of sediment to marshes has only recently become available, and such calculations should be reported in the near future (Krone, 1985).

Changes in relative mean sea levels will also change the geometries of rivers flowing into the ocean. Laboratory studies (Chang, 1967) led to the conclusion that rising sea level at the downstream end of a river will greatly enhance the river's meandering tendency. Falling sea level encourages the river to run straight and can significantly increase its sediment-carrying capacity, and thus increase estuarine sedimentation. This phenomenon was reported by geomorphologists studying water-surface fluctuations of large lakes.

Whether in an estuarial environment or not, coastal harbors will experience different effects of a change in relative mean sea level, depending on shoreline and bay bathymetry configuration. These changes can be quite complex, but estimates of them can be made with appropriate physical and/or mathematical models.

BREAKWATERS, SEA WALLS, AND JETTIES

Sea defense systems of the rubble-mound type can be easily increased in elevation by adding armor units, stone or cast concrete shapes. Normal maintenance often requires adding material to compensate for settlement and consolidation of the core materials and foundations, and adjustments for sea level rise may be part of that process. The core of a rubble-mound breakwater is usually relatively impervious. Thus, a breakwater that needs to be increased in height will require modification to the core, to the filter layers, and to the armor. In seismic regions the new design must be analyzed to assure its earthquake-resistance capacity.

Breakwaters and sea walls of solid construction, such as monolithic concrete, will be overtopped more often. Since wave damage is a function of wave height, which in turn increases in proportion to water depth, the damage may increase exponentially with sea level rise. This may require raising top levels and slopes after a period of years.

In some regions the design wave for breakwaters (and levees) is limited due to the water depth so that an increase in water depth owing to a relative increase in mean sea level will result in a larger design wave, which will require modification to the size and/or slope of the breakwater.

NAVIGATION GATES

Navigation gates are covered briefly under the section entitled "Levees." Adequate data are not available to determine where new navigation gates might be needed, and if existing gates need to be modified.

PIERS AND WHARVES

If the sea level at a pier or wharf were to rise as much as 1 ft in 50 years, and the design life is taken to be 50 years, then the deck elevation may have to be designed to be functional for both present and future conditions. One solution, for which there is precedent at the Brooklyn Navy Yard (where crane rails were built at elevations that allowed for subsidence), is to build at a level 6 in. higher than presently needed, thus splitting the difference. If wave action is expected to be significant, the underside of the deck structure may have to be kept above future wave crests, or 1 ft higher than is needed now. Piles or caissons would have to be 1 ft longer. Similarly, cranes and loading/unloading equipment would have to be designed to reach 1 ft higher above ship decks.

Piers are located on open coasts as well as harbors. The case of piers on the open coast would be more complex than discussed above. With no change in the bottom, the piers in most exposed locations would be subject to larger waves as sea level rose, because the deeper water would permit higher breakers, and thus greater wave-induced forces on structures. However, as previously discussed, changes will occur in the bottoms, and then the problem becomes quite complex.

It is possible that the change in climate, with the resulting change in numbers and intensity of storms, will be more important than a small change in relative mean sea level. For example, during the winter of 1982–1983, when an El Niño-Southern Oscillation condition existed, both the number and intensity of storms increased (Seymour et al., 1984). More than 12 ocean piers in California were either destroyed or severely damaged. It is not evident to this committee whether the numbers or intensity of storms will increase or decrease as a result of the greenhouse effect. However, by and large the expected magnitude of the consequences to coastal piers is relatively minor.

In some ports located in estuarine environments, wood pile-supported piers and wharfs are far enough upstream that the water is fresh enough to eliminate the problem of marine borers. With an increase in relative mean sea level, however, marine borers could become a problem. Jackson and Paterson (1977) reported an incident concerning the marine borer teredo that previously occurred in the San Francisco Bay area. By 1924 the worms had invaded parts of Suisun Bay and destroyed piles, thus requiring the replacement of untreated wood with creosoted timbers or even concrete structures. Before the pier at Avon could be rebuilt, it collapsed while a tanker was unloading its cargo. As the pier fell, electric light wires short-circuited, setting fire to gasoline stored in a shed on the wharf, and the pier and ship were engulfed in flames. The ship, a blazing wreck, drifted into the center of Suisun Bay. Six crew members died in the tragedy.

There may also be changes in the types and amount of marine fouling.

DRY DOCKS AND WET DOCKS

Dry docks may be adversely affected by a sea level rise, depending on the type of dock. Floating docks, ship lifts, and marine railways will merely have deeper water in which to float vessels. The latter, however, will have less clearance above high tide for repair of ships' bottoms. Graving docks, when empty, will be subjected to greater upward water pressure against the floor, if they are the gravity type. An example of this type dock was Dock No. 1 at the Long Beach Naval Shipyard, which experienced floor cracking and increased leakage in response to the subsidence of the Terminal Island area; the drainage relief system had to be altered. With a sea level rise, relieved-type graving docks may experience increased seepage and require increased pumping rates caused by the increased drainage gradient.

Wet docks, like piers and wharves, will be affected by a sea level rise depending on the type of construction. Effects will also depend on the purposes of the dock. For example, at general cargo docks, as goods are moved off a vessel, ships will gradually ride higher in the water with respect to dock level, requiring higher cranes. At container berths serviced by dock cranes, the maximum vertical travel of the cranes will clear the vessel by less and less. At bulk material docks, the vertical travel of the land-based unloader will clear objects on the ships' decks by less and less. Allowance for any increase, in all cases, will be difficult to compensate, and it may be advisable to incorporate anticipated sea level rise in the original design, despite additional cost. Fluid loading and unloading docks that depend on jointed-pipe loading arms also will need to be designed for the rising level of ships' decks and manifolds, since tankers ride higher in the water when they are empty. Loading systems that use hose connections may have more flexibility, or may be more easily modified by adjustments in hose length or supports.

HIGHWAYS, RAILROADS, BRIDGES, AND VEHICULAR TUNNELS

As sea level rises, highways and railroads across lowlands near tidal water will experience more frequent flooding during high tides and storms. This effect may be especially severe in certain estuaries where the rise in sea level will be amplified, the more so because these same estuaries are more vulnerable to storm surges as water funnels into a gradually narrowing arm of the sea. The levels of such highways and railroads may have to be raised by reballasting or adding pavement from time to time. Some highways and railroads already experience flooding during heavy rains and high tides, and these events would increase, especially at underpasses. Increased pumping capacity would be needed. Much can be learned from experiences in Japan with the rapid subsidence of some heavily populated areas that have much infrastructure of this type.

The clearance above high water will gradually diminish for bridges across water in the tidal zone. The amount of the reduction will be greater in the case of bridges upstream in estuaries where the rise of water level is amplified by funnel effects, as in Delaware Bay. Although the rise may be slow and gradual, the consequences of damage to a bridge may be so catastrophic as to warrant regular monitoring.

The clearance between the center of a suspension bridge and the water surface depends upon mean sea level, tide stage, temperature, winds, freshwater flow into an estuary, and static and dynamic loads on the bridge. An example of how elevation can be affected by structural changes in the bridge is the Golden Gate

Bridge over San Francisco Bay, California. About three decades ago a lower lateral bracing system and maintenance platform rails were installed, resulting in a downward deflection of about 3 ft from the original dead-load position at the mid-point of the center span. Recently, a lighter-weight deck element was installed. The maximum additional camber of the center span during the replacement operations was about 7 ft and occurred when the new lighter-weight deck had been placed on the Marin backspan and the mainspan, with the heavier original concrete deck on the San Francisco backspan (D. E. Mohn, Golden Gate Bridge, Highway, and Transportation District, San Francisco, California, letter, December 11, 1986).

COMMERCIAL AND INDUSTRIAL BUILDINGS

Structures near tidal water will suffer increased flooding. Surface water levels will rise, and groundwater levels, which generally are driven in part by nearby harbor and estuary levels, will follow with a time lag and amplitude that increase and decrease, respectively, with distance.

Results are likely to be increased seepage into basements, poorer storm drainage, and ponding in parking areas and approaches. The former may be countered with better waterproofing or by sump pumps and drainage; the latter may eventually require added paving.

POWER PLANTS

Most power plants on coasts use sea water for cooling, and power plants on estuaries use estuarine water for cooling. Studies should be made of the effects of rising relative mean sea level on these facilities. The "design life" of a power plant is about 40 years, a period over which mean sea level could vary from a few inches to almost 1-ft, considering the scenarios used in this report. Engineering changes required in those aspects of operation most affected by a 1-ft rise in relative mean sea level, such as cooling water intake and discharge systems, would be handled as normal changes. Environmental effects would also need to be considered.

PIPELINES

With rising sea levels, groundwater might rise above some pipelines in cities and ports, affecting corrosion rates.

Large diameter pipes are used in sewer outfalls in many coastal cities and, in some cases, for intake and discharge conduits for power plant cooling water systems. To counter the effects of sea level rise in these systems, it might be necessary to increase pumping capacity, as well as to increase the capability of outfall systems to dilute their effluent.

Another possible effect is associated with the amount of erosion that might occur on sand coasts. Pipe sections buried under the beaches and current surf zone might become exposed, while discharge ports could become covered with sand.

FLOODING AND STORM DRAINS

Flooding could be a major problem in many low-lying portions of coastal cities with a sea level rise. Examples include New Orleans, portions of which are below present sea level, and areas of San Francisco. Additional pumping will be required. More flooding would occur than at present in many areas, and levees, tide gates, and channels would be needed in some regions. Environmental conflicts would have to be examined and reconciled.

Considerable detail on planning and installation of a flood control system for the Koto delta region of Tokyo, Japan is given by Ukena et al. (1970). Because of subsidence, about 50 percent of the Koto delta area became lower than the daily low-tide level by 1953. Natural runoff from this area became nearly nonexistent, and forced drainage had to be used on a very large scale.

Subsidence in lands adjacent to the south end of San Francisco Bay between 1934 and 1967 was measured at several hundred benchmarks. About 100 mi^2 subsided more than 3 ft, with the maximum subsidence of about 8 ft. As a result, several miles of levees were raised to prevent flooding by bay waters, and flood control levees were added near the ends of the streams running into the bay (Poland, 1970). Poland states that salt water from the bay moved upstream, and channel grades crossing the subsidence area became downwarped. This resulted in sediment deposition near the stream mouths, with a reduction in channel capacity, which in turn required an increase in levee heights. Even with the higher levees, flooding occurs at times of very high runoff.

Williams (1985) points out that at present there are no standard assumptions for backwater calculations in the design of flood control projects; the usual practice is to design for the 100-year flood at a tide level of mean higher high water. An increase in relative mean sea level could result in higher flood waters upstream, with the possibility of levee overtopping. In addition, he states that the flood control systems of many areas presently rely on storage of flood waters and gravity releases at low tide. Inadequate information is available to see how a rise in relative mean sea level would affect specific areas.

In some parts of the country (for example, the New Jersey Meadows tidal marsh), mosquito control commissions have installed elaborate systems of drainage ditches and tide gates to limit backflooding of marshland when the tide rises. The flow is very sensitive to changes in mean tide level, and will be affected seriously by a rise in relative mean sea level.

La Roche and Webb (1986) estimate the cost of the expected overhaul of an urban gravity drainage system in Charleston, South Carolina for (1) present sea level, (2) an 11-in. rise, and (3) a 15-in. rise. They also estimate the cost of an overhaul if the system is built for today's sea level and later retrofitted for 11- and 15-in. rises. Presently, designing for an 11-in. rise would require larger pipes at an additional cost of $260,000 (a 5 percent increase). However, a retrofit would cost $2.4 million not including indirect costs of closing the streets.

Waddell and Blaylock (1987) conducted a similar evaluation for a watershed in Ft. Walton Beach, Florida, which is more lightly developed and employs a variety of independent measures to reduce flooding. They conclude that there are no savings to designing the system for a future rise, compared with upgrading the system if and when a rise occurs.

Titus et al. (1987) evaluate the planning implications of the preceding studies. They conclude that future sea level rise is relevant to today's design decisions where cities are overhauling urban gravity drainage systems, but not where drainage improvements are achieved by new parallel systems. In the former case, designing for a future rise is similar to insurance, with the economic merits being site specific. The report also discusses forced drainage, which was not investigated in detail by the studies.

A segment about 30 km long by 7 km wide along the coast of the Sea of Japan in Niigata and the nearby vicinity subsided by about 50 to 150 cm between 1959 and 1968 (Hirono, 1970; Takeuchi et al., 1970). Larger areas subsided by lesser amounts; most of the Niigata Plain, about 8,300 km^2, had some subsidence. Takeuchi et al. (1970) mention that in the Niigata lowland on the flood plain of the Shinano River, the subsidence decreased the ability of pumps to drain the area and also damaged the drainage canal network. They estimated the cost of reconstruction to be about 20 billion yen (then about $56 million).

HOTELS AND MALLS

The effects of a sea level rise on these facilities will be similar to those for commercial buildings, but greater emphasis will be needed to preserve the amenities that attract patronage. Remedies may be applied sooner as the effects of even a small rise in sea level become apparent. This may be especially pertinent for facilities close to the oceans. The threat of damage to a waterfront hotel, for example, might warrant extensive measures to reinforce sea walls, add beach materials, or otherwise protect the shore and hotel foundations from storm damage.

Many malls and hotels in the United States become "aged" in a few decades and undergo major and extensive renovation. It is likely that measures to accommodate small rises in relative mean sea level would be taken as a part of the renovation.

RESIDENTIAL CENTERS

As more and more people move to shore housing sites, a rise in sea level will become more evident to greater numbers of people. The effects will be most noticeable in beach erosion, sea wall damage, and flooding of lower levels, drives, and swales. Since many such areas are also subject to subsidence from earth compaction, groundwater pumping, or tectonic movement, the effect of the rise may be accentuated.

For example, in the area around Baytown, Texas, on Galveston Bay, subsidence has caused frequent high-tide flooding of land that slopes toward the bay at gradients of 1 ft/mi, on the average.

WATER SUPPLY SYSTEMS

A number of effects on water supply systems (particularly on water quality) can be realized with changes in relative mean sea level. Some of these are considered below for both groundwater and surface-water sources.

Perched Fresh Water

The increase in hydrostatic pressure with depth is about 3 percent greater in sea water than in fresh water. If the surface of the freshwater table is X feet above mean sea level, under static conditions, the freshwater-seawater interface would be about 40X ft below mean sea level.

As sea level rises, this "bubble" of fresh water should simply float in the salt water at an elevation that is higher by the amount of sea level rise. Thus, no significant effects are expected. Additionally, in many locations on barrier spits and islands the water supply is presently brought from other regions and thus a rise in sea level should present no additional problem.

Aquifers

Coastal aquifers normally flow toward adjacent surface waters such as lakes, rivers, estuaries, or the sea. Excessive pumping for irrigation and municipal water supplies can reverse the flow so that water is recharged to the aquifer. If such recharge occurs near the mouth of a river, a rise in sea level can recharge the aquifer with sea water. Conditions are aggravated during droughts, when the saltwater wedge advances upstream and when pumping for irrigation is augmented. An example of a potential problem is the Delaware River, which recharges the Potomac-Raritan-Magothy aquifer (the source of water for many wells in New Jersey) above river mile 98 (Hull and Titus, 1986; Camp, Dresser and McKee, Inc., 1982; Hull and Tortoriello, 1979).

Some possible engineering responses to this problem are: (1) modifying the elevation of the aquifer's connection to the estuary to reduce the landward penetration of the salt wedge; (2) reducing the permeability of the sediment where the aquifer communicates with the estuary to reduce the rate of seawater recharge; and (3) increasing recharge during periods of high precipitation.

Freshwater Intakes from Upstream Regions of Estuaries

A rise in relative mean sea level could have far-reaching effects on taking fresh water from upstream regions in estuaries. As an example, consider the Sacramento-San Joaquin delta region of California from the standpoint of maintaining water quality standards. The California Water Resources Control Board (1983 and 1984) has stated:

> The Delta is a vital link between river systems of the Sacramento Valley and the water deficient areas to the south and west of the Delta. Two major systems—the State Water Project (SWP) operated by the Department of Water Resources (Department) and the federal Central Valley Project (CVP) operated by the United States Bureau of Reclamation (Bureau)—withdraw supplies from the Delta for use in areas of need. These projects are the two largest water diversions from the Delta. They provide municipal supplies to areas where over 14 million people live and support an extremely productive agricultural economy in the San Joaquin Valley.
>
> The underlying principle of these standards is that water quality in the Delta should be at least as good as those levels which would have been available had the state and federal projects not been constructed, as limited by the constitutional mandate of reasonable use. The standards include adjustments in the levels of protection to reflect changes in hydrologic conditions experienced under different water year types.

In a recent decision of a state court of appeal, it was ruled that the California Water Resources Control Board has "compromised its important water quality role by defining its role too narrowly in terms of enforceable water rights" (Milstein, 1986).

A possible rise of mean sea level of between 0.5 and 1.5 m by the year 2100 could have an important impact on the structures and methods necessary to maintain water quality standards. According to Kennedy (1986), there are many alternatives to transferring fresh water through and around the delta; several are described by the California Department of Water Resources (1983a).

In the *New York Times* (March 21, 1986), a newly developed scheme was announced to liberate New York City from the periodic water shortages that result from its dependence on the drought-prone tributaries of the Delaware River. An existing pump station on the Hudson River at Chelsea, on the east bank near Poughkeepsie, would be enlarged and integrated into a plan of new reservoir construction. During high-discharge stages of the river, the fresh water would be "skimmed" and put in the storage reservoirs. At low-discharge times the salt water comes dangerously close to the pump station. At present, the Hudson River is tidal as far as Albany. A sea level rise of 0.5 m would bring the saltwater wedge above the level of Poughkeepsie.

Hull and Titus (1986) estimate an increased salinity of the Delaware estuary during a repeat of the 1960s drought for 2.4-ft and 8.2-ft rises in sea level. A 2.4-ft rise would cause the 250-ppm isochlor (the "salt front") to move 7 miles upstream, on the average. During 15 percent of the tidal cycles, the river at Philadelphia's Torresdale drinking water intake would have elevated sodium concentrations exceeding 50 ppm (the New Jersey drinking water standard) if no countermeasures were taken. Lennon et al. in Hull and Titus (1986) conclude that the elevated salt levels in the river could contaminate parts of the Potomac-Raritan-Magothy aquifer, which is pumped at a point below sea level and recharged by the river. The report cites several mitigation measures, such as new reservoirs, and recommends long-term planning for consequences of sea level rise as well as possible changes in drought frequency caused by the greenhouse effect.

LANDFILLS AND WASTE DISPOSAL SITES

A rise in sea level can affect landfills and disposal sites in two ways: (1) direct overtopping and erosion, or (2) changes in the level of the aquifer and the groundwater leaching pattern. Dikes, similar to those currently used in some containment areas, could be designed and constructed to counter both of these effects.

OFFSHORE PLATFORMS AND ARTIFICIAL ISLANDS

The productive lives of offshore platforms and artificial islands used in the production of oil and gas is on the order of 25 years. They should not be affected very much by eustatic rise in mean sea level in that time span. The problem of relative rise in mean sea level owing to subsidence can be of much greater importance. An article in *News of Norway* (May 22, 1985) describes how the Ekofisk platforms have sunk 2 m (6 ft) and continue to subside at a rate of 1–4 cm/mo or 12–48 cm/yr (about 5–9 in.). The article mentions a conclusion in a recent report by the Norwegian classification society, Det Norske Veritas, that several platforms on the field would be total losses in the event of a major storm.

An article in the *Financial Times* of London (April 11, 1986) states that a massive rescue plan has been proposed to raise six steel oil platforms by 6 m. It claims the platforms have sunk nearly 3 m because of the weight of 3,000 m of rock overlying the field's oil and gas reservoir. Subsidence due to the removal of hydrocarbons is also a likely contributing factor. A report commissioned by Phillips Petroleum to study two of the peripheral platforms in the field concluded that total loss, given a 100-year wave, could not be ruled out if the platforms sank 8.5 ft, and even a heavy storm would cause serious damage. A number of alternatives were considered. The lowest cost alternative, and one which could be done with only 18 days production shutdown, would cost $286 million (Anonymous, 1986). The plan would consist of jacking up the decks of the platforms and installing extensions. The work is expected to start in June 1987.

8
Decisions for the Future

STRATEGIC DECISIONS

An accelerated rise in relative sea level would force people who live on the coasts to face a number of important decisions. In the past, keeping the coastal infrastructure above the historical slow rise in sea level for the most part has been achieved through normal maintenance or abandonment of facilities. The effects from sea level rise scenarios adopted in this report imply that a more considered and planned approach to the preservation or abandonment of coastal facilities and communities is needed.

As discussed in Chapters 3–7, the types of responses to sea level rise include retreat from the shoreline or the use of structures to prevent flooding and shoreline recession. The choice of response strategy will depend on several factors. A clear understanding of the natural processes underlying shoreline erosion and knowledge of the efficacy of coastal structures are important to both strategies. Additional site-specific considerations are the economics involved as well as the social and environmental costs. Limited information exists concerning these issues.

For well-developed coastal communities, with a high density of buildings and expensive shoreline facilities such as harbors and resorts, the strategy of choice in all likelihood will be to protect the existing infrastructure. For eroding shorelines that are less developed the decision becomes more difficult. Because the costs and benefits of protection must be weighed against those of retreating from the shoreline, consideration must be given to economic, social, environmental, and geological and geographical factors.

Economic Factors

Long-Term Costs

The initial cost to stabilize a shoreline is expensive, whether it is through beach fill, groins, or sea walls. Continual maintenance of the structures represents an ongoing cost to the community; the magnitude of the cost will vary with time, depending on future sea levels. However, if these costs are outweighed by the benefits of maintaining the coastal infrastructure and beach, then stabilization is the rational choice.

Stabilization in the low wave-energy environments of most coastal bays and wetlands will be much less expensive than in open-coast areas. Social and environmental factors must be considered when making decisions about wetlands.

The decision to retreat may seem attractive on first consideration because the construction cost for engineering structures is zero. However, retreat involves foregoing the use of land and perhaps buildings, which represents a large short- to medium-term loss, which might be outweighted by the long-term cost-saving of doing nothing in an engineering context.

Design Life Versus Remedial Measures

In planning for sea level rise, it is necessary to consider whether a stabilization project needs to be made at the outset, or whether remedial measures can be taken periodically during the life of the structure or facility. Two examples illustrate the difference. A highway is to be built across low-lying land to an estuary; its useful life might be on the order of 60 years. For every foot of increased elevation (to allow for the sea level rise) the cost may increase by millions of dollars. On the other hand, with no absolute assurances of the rate of future sea level rise, and since repaving and raising the road to counteract the expected local subsidence will be undertaken anyway, the logical decision is probably to deal with sea level as part of ongoing maintenance rather than building for it initially.

A second example concerns the construction of a beach-front park and pavilion with expensive shops. A rise in sea level of 1 ft may jeopardize the entire investment, which may also have a 60-year design life. In this case, the prudent design might include an extra foot of elevation and a horizontal setback of X feet to safeguard against expensive reconstruction.

Social Factors

Degree of Risk from Sea Level Rise

Along with other factors, the consequences of damage from a sea level rise in relation to the degree of risk should be evaluated. If risk to human life is a factor, as in the case of a beach-front hotel, then a higher factor of safety is justified, requiring planning for sea level rise in the initial design.

An example of a structure with a high risk to life would be a levee protecting a major population center. The consequences of levee overtopping, failure, and flooding to a large urban area would require careful and thorough analysis and a conservative allowance for sea level rise, including the wave action accompanying major storms.

Maintenance Capability

Waterfront facilities are owned by many individuals and agencies who have varying attitudes and capabilities with regard to funding, monitoring, and maintaining their properties. Facilities that are likely to remain unattended for long periods, or those whose owners are unable to modify or maintain them, may need to be built with a conservative allowance for sea level rise. In this category would be beach-front hotels, graving docks, and high-level bridges. On the other hand, owners of levees, roadways, breakwaters, and similar facilities are likely to monitor their structures, and are in a position to adjust them to withstand the effects associated with rising sea levels.

Further, once the decision is made to stabilize, it becomes more difficult in the future to change the decision. With the retreat option or a decision to do nothing, the cost to reverse the decision if the economics become favorable for stabilization is far less.

ENVIRONMENTAL FACTORS

The environmental effects of coastal structures or retreat from the shoreline vary with the site. It is clear that the use of coastal structures may safeguard breeding areas or specialized habitats, or result in their loss. The same is true of the retreat alternative.

For coastal wetlands located in relatively protected regions, the costs of stabilizing shoreline positions would probably not be high, due to the less energetic wave climate in bays and lagoons. If there is a retreat along bay shorelines, or a ban on shoreline stabilization, marshlands will be allowed to retreat with the sea level rise.

Geological/Geographical Factors

The location of the site of interest is extremely important in the decision-making process. High-energy shorelines are more expensive to maintain than low-energy shorelines. Coastlines with an abundance of sand in the littoral stream are more easily maintained than those that exist on shorelines with little or no sand supply.

Local factors can be the predominant cause of shoreline recession, overwhelming present day erosion rates attributable to sea level rise. For example, tidal inlets trap sand from the littoral transport, and the beaches on the downdrift side of inlets are often in a state of erosion. The most effective solution for affected communities is to solve first the more massive and immediate local erosion problem, by bypassing sand around the inlet, and then to address the long-term sea level rise problem.

The concept of sand rights should be involved in the decision-making process. The impacts of the actions of one community on neighboring communities that share the same sand supply system must be evaluated. For example, if eroding bluffs along the shoreline are known to be the source of sand for downdrift communities, then decisions to stabilize the bluffs will have far-reaching consequences. In general, sources of sand should be protected to allow the sand to move in the littoral system, while sinks of sand are locations where stabilization would be extremely effective. If structural means are employed that reduce the natural sand supply, consideration should be given to requiring mitigation measures, such as supplying sand from an alternative source. The ideal location for communities that need to stabilize the beaches would be at the terminus of the sand supply for a given coastal reach.

NATIONAL POLICY DECISIONS

Responding to changing sea level will take on greater urgency and relevance in the future as several apparently irreversible trends combine to render the coastal zone more hazardous as a site for major investment. These trends include

- increasing human settlement, services, and installation in coastal areas susceptible to inundation, erosion, or destruction due to increases in relative sea level, storm surges, and tidal crescendos;
- land subsidence, which affects the U.S. East Coast from Long Island to Georgia, the Gulf Coast from Texas to Louisiana, and local areas on the Pacific Coast; and
- the steady increase of atmospheric greenhouse gases, which is theorized to enhance global warming, thus contributing to glacial melting and eustatic sea level rise.

Coping with these trends will require

- research, data acquisition, and analysis of the specific effects of sea level rise in relation to other environmental changes and the response of specific coastal works to rise; and
- review and coordination of national and regional policy concerning the coastline of the United States and its dependencies, especially to take account of new knowledge.

Prediction of climatic, oceanographic, and geologic processes that are potentially hazardous to coastal structures within 10 to 50 years and the ability to warn the public of hazards are also needed. The expertise for developing mechanisms to cope with long-term sea level rise is available from specialists in these areas: coastal surveying, monitoring, and preservation; coastal and harbor engineering; tidal measurement and prediction; meteorology; climatology; geodesy; geology, oceanography; ecology; and coastal management.

9
Conclusions and Recommendations

CONCLUSIONS

1. Relative mean sea level, on statistical average, is rising at the majority of tide gauge stations situated on continental coasts around the world. Relative mean sea level is generally falling near geological plate boundaries and in formerly glaciated areas such as Alaska, Canada, Scandinavia, and Scotland. Relative mean sea level is not rising in limited areas of the continental United States, including portions of the Pacific Coast.

2. The contrasting signals concerning relative mean sea level behavior in different parts of the United States (and the world in general) are interpreted as due to differing rates of vertical motion of the land surfaces. Subsidence and glacial rebound are significant contributors to vertical land displacements.

3. Large, short-term (2–7 year) fluctuations worldwide are related to meteorological phenomena, notably shifts in the mean jet-stream path and the El Niño-Southern Oscillation mechanisms, which lead to atmospheric pressure anomalies and temperature changes that may cause rise or fall of mean sea level by 15–30 cm over a few years.

4. Studies of a very small number of tide gauge records dating more than 100 years (the oldest being Amsterdam, started in 1682)

show that after removal of the subsidence factor where known, mean sea level has been fluctuating through a range of not more than 40–150 cm (in long-term fluctuations) for at least 300 years.

5. The geological record over the last 6,000 years or so indicates that there has been a general, long-term rise with short-term fluctuations probably not exceeding 200 cm during the last 1,500 years.

6. Monitoring of relative mean sea level behavior is at present inadequate for measuring the possible global result of future climate warming due to rising greenhouse gases. The most serious gaps in present tide gauge coverage are in three areas: (a) high polar latitudes, (b) midoceanic locations, and (c) the entire Southern Hemisphere.

7. Because of localization of many extreme subsidence processes, especially those connected with anthropogenic extraction of fluids such as groundwater and hydrocarbons, tide gauges are needed at every major coastal city to gather data to assist in evaluating the long-term regional trend of relative mean sea level.

8. The risk of accelerated mean sea level rise is sufficiently established to warrant consideration in the planning and design of coastal facilities. Although there is substantial local variability and statistical uncertainty, average relative sea level over the past century appears to have risen about 30 cm relative to the East Coast of the United States and 11 cm along the West Coast, excluding Alaska, where glacial rebound has resulted in a lowering of relative sea level. Rates of relative sea level rise along the Gulf Coast are highly variable, ranging from a high of more than 100 cm/century in parts of the Mississippi delta plain to a low of less than 20 cm/century along Florida's west coast.

9. Accelerated sea level rise would clearly contribute toward a tendency for exacerbated beach erosion. However, in some areas, anthropogenic effects, particularly in the form of poor sand management practices at channel entrances, constructed or modified for navigational purposes, have resulted in augmented erosion rates that are clearly much greater than would naturally occur. Thus, for some years into the future, sea level rise may play a secondary role in these areas.

10. As noted previously, the two response options to sea level rise are stabilization and retreat. Retreat is most appropriate in areas with a low degree of development. Given that a "proper"

choice exists for each location, selecting an incorrect response alternative could be unduly expensive.

11. There does not now appear to be reason for emergency action regarding engineering structures to mitigate the effects of anticipated increases in future eustatic sea level rise. Sea level change during the design service life should be considered along with other factors, but it does not present such essentially new problems as to require new techniques of analysis. The effects of sea level rise can be accommodated during maintenance periods or upon redesign and replacement of most existing structures and facilities. There are very limited geographic areas where current subsidence rates may require near-term action as has been the case in Japan and Terminal Island, California.

12. When not restrained by funding, availability of materials, or work force, construction of almost any conceivable protection against sea level rise can be carried out in a very short time; short, that is, relative to the rate of sea level rise.

13. Defensive or mitigative strategies are site specific and cannot be developed nationwide on the basis of a blanket generalization or comprehensive legislation.

RECOMMENDATIONS

1. The prognosis for sea level rise should not be a cause for alarm or complacency. Present decisions should not be based on a particular sea level rise scenario. Rather, those charged with planning or design responsibilities in the coastal zone should be aware of and sensitized to the probabilities of and quantitative uncertainties related to future sea level rise. Options should be kept open to enable the most appropriate response to future changes in the rate of sea level rise. Long-term planning and policy development should explicitly consider the high probability of future increased rates of sea level rise.

2. The three previously described scenarios of sea level rise used in this study (see Figure 2-2) provide a useful range of possible future sea level changes for design calculations. The general shape of these curves is concave upward with greater rates of rise in the distant future than those in the next decade or so. The confidence that these scenarios will encompass the actual levels decreases with increasing time, and significant deviations outside the range of these scenarios are possible, including an amelioration in the rate

of rise. Thus, the committee recommends that these projections be updated approximately every decade to incorporate additional data and to provide an improved basis for planning and response to the rise.

3. Practitioners can more readily incorporate the implications of sea level rise if probabilities reflecting uncertainties are attached to the projections. Thus, it is recommended that appropriate statistical techniques be applied to develop a probability distribution associated with sea level rise through the year 2100 and that all updated projections include such information.

4. Feasibility studies for coastal projects (e.g., shore protection projects of the U.S. Army Corps of Engineers and storm surge studies of the Federal Emergency Management Agency) should consider the high probability of accelerated sea level rise. It may be some time before precise estimates of future sea level rise are possible. In the meantime, the risks associated with a substantial rise should not be disregarded. Instead, feasibility studies should consider which designs are most appropriate for a range of possible future rates of rise. Strategies that would be appropriate for the entire range of uncertainty should receive preference over those that would be optimal for a particular rate of rise but unsuccessful for other possible outcomes.

5. The federal government should acquire long-term reliable accurate data from a water-level measuring system for open-ocean stations at scientifically important locations throughout the world. Critical stations should include documentation of vertical ground motion and the temporal salinity and temperatures of the water column. Tide gauges should be installed at major coastal cities.

6. The important decision for maintaining or abandoning coastal facilities in the face of rising sea level should be well documented by scientific knowledge. Agencies that fund coastal research, such as the U.S. Navy, U.S. Army, National Science Foundation, National Oceanic and Atmospheric Administration, U.S. Geological Survey, and the Environmental Protection Agency, should increase their funding for coastal processes research. The federal research funding effort should focus on studies directed toward understanding nature's response to relative sea level rise and developing appropriate engineering responses. A substantial portion of this research should be conducted at universities and other laboratories and centers throughout the coastal United States to ensure the development of requisite engineering capability in regions of the country where it will be most helpful.

References

Allen, D. R., and M. N. Mayrega. 1970. The mechanics of compaction and rebound, Wilmington Oil Field, Long Beach, California. Pp. 410–423 *in* Land Subsidence: Proceedings of the Tokyo Symposium, September 1969, Vol. II. Paris: UNESCO.

Anonymous. 1986. The Five-Platform Lift. Noroil 14(8):130–133.

Aubrey, D. G., and K. O. Emery. 1983. Eigenanalysis of recent United States sea levels. Continental Shelf Research 2:21–33.

Bakker, J. P. 1981. Transgression phases and the frequency of storm floods in the Netherlands in recorded time. Pp. 51–56 *in* Overwash Processes, Benchmark Papers in Geology, V. 58, S. Leatherman, ed. New York: Van Nostrand Reinhold.

Barnett, T. P. 1983a. Recent changes in sea level and their possible causes. Climate Change 5:15–38.

Barnett, T. P. 1983b. Long-term changes in dynamic height. J. Geophys. Res. 88(C14):9547–9552.

Barth, M. C., and J. G. Titus. 1984. Greenhouse Effect and Sea Level Rise. New York: Van Nostrand Reinhold. 325 pp.

Bascom, W. H. 1951. The relationship between sand size and beach-face slope. Trans. Am. Geophys. U. 32(6):866–874.

Bindschadler, R. A. 1985. Contribution of the Greenland Ice Cap to changing sea level: Present and future. *In* Glaciers, Ice Sheets, and Sea Level. Washington, D.C.: National Academy Press.

Bird, E. C. F. 1985. Coastline Changes—A Global Review. Chichester: J. Wiley–Interscience. 219 pp.

Bird, E. C. F. 1976. Shoreline changes during the past century. Proceedings of the 23rd International Geographical Congress, Moscow. Elmsford, N.Y.: Pergamon. 54 pp.

Bruun, P. 1983. Review of conditions for uses of the Bruun rule of erosion. Coastal Eng. 7:77–89.

Bruun, P. 1962. Sea-level rise as a cause of shore erosion. J. Waterways and Harbors Div. ASCE 88:117–130.

Bruun, P. 1954. Coastal Erosion and Development of Beach Profiles, Beach Erosion Board Technical Memorandum No. 44. Washington, D.C.: U.S. Army Corps of Engineers.

Bruun, P., and F. Gerritsen. 1960. Stability of Coastal Inlets. Amsterdam: North Holland Publishing Company.

California Department of Water Resources. 1983a. Alternatives for Delta Water Transfer. Sacramento: Department of Water Resources. 70 pp.

California Department of Water Resources. 1983b. Flood Hazard Mitigation Plan for the Sacramento-San Joaquin Delta. Sacramento: Department of Water Resources. 21 pp.

California Water Resources Control Board. November 1983 and February 1984. Water Right Decision 1594 and Order WR 84-2. Sacramento: Department of Water Resources. 61 pp.

Camp, Dresser and McKee, Inc. 1982. Groundwater Management Plan for Study Area 1: Coastal Plain Formations. Prepared for Delaware River Basin Commission, West Trenton, New Jersey.

Cartwright, D. E., T. P. Barnett, C. J. R. Garrett, W. E. Carter, R. Pelletier, T. Pyle, and K. R. Thompson, eds. 1985. Changes in Relative Mean Sea Level. IAPSO Advisory Committee on Tides and Mean Sea Level. EOS (Nov. 5): 754–756.

Chabreck, R. H. 1972. Vegetation, Water and Soil Characteristics of the Louisiana Coastal Region. Louisiana Agricultural Experiment Station Bull.

Chang, H. 1967. Hydraulics of Rivers and Deltas. Ph.D. dissertation. Colorado State University, Fort Collins.

Dalrymple, R. A., R. B. Biggs, R. G. Dean, and H. Wang, 1986. Bluff recession rates in Chesapeake Bay. J. Waterway, Port, Coastal, and Ocean Eng. 112(1):164–168.

Davies, J. L. 1957. The importance of cut and fill in the development of beach sand ridges. Australian J. Sci. 20:105–111.

Davis, A. B., Jr. Undated. Galveston's bulwark against the Sea: History and the Galveston seawall. Galveston: U.S. Army Corps of Engineers.

Davis, A. B., Jr. 1961. The Galveston sea wall. Shore and Beach 29(2):6–37.

Dean, R. G. 1983. Shoreline Erosion Due to Extreme Storms and Sea-Level Rise. Department of Coastal and Oceanographic Engineering. Gainesville: University of Florida. 58 pp.

Dean, R. G. 1977. Equilibrium Beach Profiles: U.S. Atlantic and Gulf Coasts. Department of Civil Engineering Report No. 12. Newark: University of Delaware. 45 pp.

Dean, R. G. 1976. Beach erosion: Causes, processes, and remedial measures. Pp. 259–296 *in* CRC Critical Reviews in Environmental Control. Boca Raton, Fla.: CRC Press.

Dean, R. G., and R. A. Dalrymple. 1984. Water Wave Mechanics for Engineers and Scientists. Englewood Cliffs, N.J.: Prentice-Hall, Inc. 353 pp.

Dean, R. G., and E. M. Maurmeyer. 1983. Models for beach profile response. Pp. 151–165 *in* CRC Handbook of Coastal Processes and Erosion Control. Boca Raton, Fla.: CRC Press.

DeLaune, R. D., R. H. Baumann, and J. G. Gosselink. 1983. Relationships among vertical accretion, coastal submergence, and erosion in a Louisiana Gulf Coast marsh. J. Sed. Petrol. 53:147–157.

Demarest, J. M., and S. P. Leatherman. 1985. Mainland influence on coastal transgression: Delmarva Peninsula. Mar. Geol. 63:19–33.

Dietz, R. S. 1963. Wave-base, marine profiles of equilibrium, and wave-built terraces: A critical appraisal. Geol. Soc. Am. Bull. 74:971–990.

Emery, K. O. 1980. Relative sea levels from tide gauge records. Proc. Nat. Acad. Sci. 77:6968–6972.

Engineering News. August 26, 1915. 74(9):424–426.

Engineering News. April 24, 1902. 47(17):343–344.

Evans, M. W., A. C. Hine, D. F. Belknap, and R. A. Davis. 1985. Bedrock controls on barrier island development: West central Florida coast. Mar. Geol. 63:263–283.

Everts, C. 1985. Effects of sea level rise and net sand volume change on shoreline position at Ocean City, Maryland. Pp. 67–98 *in* Potential Impacts of Sea Level Rise on the Beach at Ocean City, Maryland, J. G. Titus, ed. Washington, D.C.: U.S. Environmental Protection Agency.

Everts, C. H., J. P. Battley, and P. N. Gibson. 1983. Shoreline Movements Between Cape Henry, Virginia and Cape Hatteras, North Carolina, 1949–1980. WES-CERC Technical Report 83-1. Vicksburg, Miss.: U.S. Army Engineer Waterways Experiment Station. 111 pp.

Fairbridge, R. W. 1961. Eustatic changes in sea level. Physics and Chemistry of the Earth 4:99–185.

Fairbridge, R. W., and O. A. Krebs. 1962. Sea level and the Southern Oscillation. Geophys. J., Royal Astron. Proc. 6(4):532–545.

Fairbridge, R. W., and W. S. Newman. 1968. Postglacial coastal subsidence of the New York area. Zeitschr. f. Geomorph., N. F. 12(3):296–317.

Fenneman, M. M. 1902. Development of the profile of equilibrium of the subaqueous shore terrace. J. Geol. 6(4):532–545.

Field, M. E., and D. B. Duane. 1976. Post-Pleistocene history of the United States inner continental shelf: Significance to origin of barrier islands. Geol. Soc. Am. Bull. 87:691–702.

Florida, State Department of Natural Resources. 1986. A proposed beach management program for the state of Florida. Tallahassee: Department of Natural Resources.

Gilbert, G. K. 1917. Hydraulic Mining Debris in Sierra Nevada. Professional paper 105. Washington, D.C: U.S. Geological Survey.

Goldberg, E. D., J. J. Griffin, V. Hodge, M. Koide, and H. Windom. 1979. Pollution history of the Savannah River estuary. Environ. Sci. Tech. 13:588–594.

Gornitz, V., S. Lebedeff, and J. Hansen. 1982. Global sea level trends in the past century. Science 215:1611–1614.

Gutenberg, B. 1941. Changes in sea level, postglacial uplift and mobility of the earth's interior. Bull. Geol. Soc. Am. 52:721–772.

Hallermeier, R. J. 1981. A profile zonation for seasonal sand beaches from wave climate. Coastal Eng. 4:253–277.

Hands, E. B. 1983. The Great Lakes as a test model for profile response to sea level changes. Pp. 167–189 *in* CRC Handbook of Coastal Processes and Erosion Control. Boca Raton, Fla.: CRC Press.

Hands, E. G. 1981. Predicting Adjustments in Shore and Offshore Sand Profiles on the Great Lakes. CETA 81-4. Vicksburg, Miss.: Coastal Engineering Research Center. 25 pp.

Hands, E. B. 1976. Observations of Barred Coastal Profiles Under the Influence of Rising Water Levels, Eastern Lake Michigan, 1967–1971. TR-76-1. Vicksburg, Miss.: Coastal Engineering Research Center.

Hawaii Coastal Zone Management Program. 1985. Effects on Hawaii of a Worldwide Rise in Sea Level Induced by the Greenhouse Effect. Prepared in response to Senate Resolution 137, 1984. Honolulu: Department of Planning and Economic Development. 10 pp.

Hayes, M. O. 1979. Barrier island morphology as a function of tidal and wave regime. Pp. 1–27 *in* Barrier Islands, S. P. Leatherman, ed. New York: Academic Press.

Hayes, M. O. 1967. Hurricanes as geological agents, South Texas coast. Am. Assoc. Petrol. Geol. Bull. 51:937–942.

Herbich, J. B. 1975. Coastal and Deep Ocean Dredging. Houston: Gulf Publishing.

Hicks, S. D. 1984. Inside vs. outside tide stations for the measurement of sea level trends. Paper presented at the Barrier Island Workshop of the U.S. Army Corps of Engineers, Nags Head, North Carolina. November.

Hicks, S. D. 1978. An average geopotential sea level series for the United States. J. Geophys. Res. 83(C3):1377–1379.

Hicks, S. D., H. A. Debaugh, Jr., and L. E. Hickman, Jr. 1983. Sea Level Variations for the United States, 1955–1980. Rockville, Md.: National Oceanic and Atmospheric Administration. 170 pp.

Hirono, T. 1970. Niigata ground subsidence and groundwater changes. Pp. 144–161 *in* Land Subsidence: Proceedings of the Tokyo Symposium, September, 1969. Paris: UNESCO.

Hoese, H. D. 1967. Effect of higher than normal salinities on salt marshes. Contrib. Mar. Sci. 12:249–261.

Hoffman, J. S., D. Keyes, and J. G. Titus. 1983. Projecting Future Sea Level Rise; Methodology, Estimates to the Year 2100, and Research Needs. Washington D.C.: U.S. Environmental Protection Agency. 121 pp.

Hoffman, J. S., J. B. Wells, and J. G. Titus. 1986. Future global warming and sea level rise. *In* Iceland Coastal and River Symposium '85, G. Sigbjarnarson, ed. Reykjavik: National Energy Authority.

Howard, J. D., O. H. Pilkey, and W. Kaufman. 1985. Strategy for beach preservation proposed. Geotimes 30(12):15–19.

Hoyt, J. H., and V. J. Henry. 1967. Influence of island migration on barrier island sedimentation. Geol. Soc. Am. Bull. 78:77–86.

Hudson, W. S. 1905. The great sea wall at Galveston. Scientific American (August 26):163–164.

Hull, C. H. J., and R. C. Tortoriello. 1979. Sea level trend and salinity in the Delaware estuary. West Trenton, N.J.: Delaware River Basin Commission.

Hull, C. H. J., and J. G. Titus, eds. 1986. Greenhouse Effect, Sea Level Rise, and Salinity in the Delaware Estuary. Trenton, N.J.: Delaware River Basin Commission and U.S. Environmental Protection Agency. 88 pp.

Inman, D. L., and C. E. Nordstrom. 1971. On the tectonic and morphologic classification of coasts. J. Geol. 79:1–21.

Jackson, W. T., and A. M. Paterson. 1977. The Sacramento-San Joaquin Delta—The Evolution and Implementation of Water Policy: An Historical Perspective. Contribution No. 163. California Water Resources Center, University of California, Davis.

Johnson, D. 1929. Studies of Mean Sea Level, Bulletin No. 70. Washington, D.C.: National Academy Press.

Kana, T. W., J. Michel, M. O. Hayes, and J. R. Jensen. 1984. The physical impact of sea-level rise in the area of Charleston, S.C. Pp. 105–150 *in* Greenhouse Effect and Sea Level Rise, M. C. Barth and J. G. Titus, eds. New York: Van Nostrand Reinhold.

Kennedy, D. N. 1986. Report of Activities of the Department of Water Resources. Sacramento: California Department of Water Resources. 29 pp.

Komar, P. D. 1976. Beach Processes and Sedimentation. Englewood Cliffs, N.J.: Prentice-Hall. 429 pp.

Kraft, J. C. 1971. Sedimentary facies patterns and geologic history of a Holocene marine transgression. Geol. Soc. Am. Bull. 82: 2131–2158.

Kriebel, D. L., and R. G. Dean. 1985. Estimates of Erosion and Mitigation Requirements Under Various Scenarios of Sea Level Rise and Storm Frequency for Ocean City, Maryland. Washington, D.C.: U.S. Environmental Protection Agency.

Krone, R. B. 1985. Simulation of marsh growth under rising sea level. Pp. 13–16 *in* Hydraulics and Hydrology in the Small Computer Age. ASCE Hydraulics Division Speciality Conference, Orlando, Florida. August.

Krone, R. B. 1979. Sedimentation in the San Francisco Bay system. Pp. 85–96 *in* San Francisco Bay, The Urbanized Estuary, T. J. Connors, ed. Washington: AAAS, Pacific Division.

Kuenen, Ph. H. 1950. Marine Geology. New York: John Wiley & Sons.

Kuhn, G. G., and F. P. Shepard. 1981. Should Southern California build defenses against violent storms resulting in lowland flooding as discovered in records of past century. Shore and Beach 49(4):3–11.

Lamb, H. H. 1984. Climate in the last thousand years: Natural climatic fluctuations and change. Pp. 25–64 *in* The Climate of Europe: Past, Present and Future, H. Flohn and R. Fantechi, eds. Dordrecht: D. Reidel.

LaRoche, T. B., and M. K. Webb. 1986. Impact of accelerated sea level rise on drainage systems in Charleston, S.C. *In* Potential Impacts of Sea Level Rise on Coastal Drainage Systems. Washington, D.C.: U.S. Environmental Protection Agency.

Leatherman, S. P. 1985. Geomorphic effects of accelerated sea-level rise on Ocean City, Maryland. *In* Potential Impacts of Sea Level Rise on the Beach at Ocean City, Maryland, J. G. Titus, ed. Washington, D.C.: U.S. Environmental Protection Agency. 34 pp.

Leatherman, S. P. 1984a. Shoreline evolution of North Assateague Island, Maryland. Shore and Beach 52:3–10.

Leatherman, S. P. 1984b. Coastal geomorphic responses to sea-level rise: Galveston Bay, Texas. Pp. 151–178 *in* Greenhouse Effect and Sea Level Rise, M. C. Barth and J. G. Titus, eds. New York: Van Nostrand Reinhold.

Leatherman, S. P. 1979a. Migration of Assateague Island, Maryland, by inlet and overwash processes. Geology 7(104):1–7.

Leatherman, S. P. 1979b. Environmental Geologic Guide to Cape Cod National Seashore. SEPM-ES Special Publication. 249 pp.

Leatherman, S. P., T. E. Rice, and V. Goldsmith. 1982. Virginia barrier island configuration: A reappraisal. Science 215:285–287.

Letzsch, S. W., and R. W. Frey. 1980. Deposition and erosion in a Holocene tidal salt marsh, Sapelo Island, Georgia. J. Sed. Petrol. 50:529–542.

Lisitzin, E. 1974. Sea-Level Changes. Amsterdam, N.Y.: Elsevier.

Lisitzin, E. 1958. Sea Level Changes. Elsevier Oceanography Series 8. New York: Elsevier.

May, S. K., R. Dolan, and B. P. Hayden. 1983. Erosion of U.S. shorelines. EOS 64:551–552.

McGowen, J. H., L. E. Garnev, and B. H. Wilkinson. 1977. The Gulf shoreline of Texas: Processes, characteristics, and factors to use. Texas Bureau of Economic Geology Circ. 77-3. 27 pp.

Mehta, A., and R. Philip. 1986. Bay Superelevation: Causes and Significance in Coastal Water Level Response. Gainesville: University of Florida. (Report to the Committee on Engineering Implications of Changes in Relative Mean Sea Level.)

Meier, M. F. 1984.Contribution of small glaciers to global sea level. Science 226(4681):1418–1421.

Milstein, S. 1986. Historic court ruling on state water quality. San Francisco Chronicle, May 29, 1986.

Moore, B. 1982. Beach Profile Evolution in Response to Changes in Water Level and Wave Height. Newark: University of Delaware.

Munk, W., R. Revelle, P. Worcester, and M. Zumberge. 1985. Strategy for Future Measurements of Sea Level. La Jolla, Calif.: Scripps Institution of Oceanography. Unpublished manuscript.

Murayama, S. 1970. Land Subsidence in Osaka. Pp. 105–129 *in* Land Subsidence: Proceedings of the Tokyo Symposium, September 1969, Vol. I. Paris: UNESCO.

National Research Council, Committee on National Dredging Issues. 1985a. Background Paper of the Technical Panel on Ports, Harbors, and Navigation Channels of the Marine Board. Unpublished manuscript.

National Research Council, Polar Research Board. 1985b. Glaciers, Ice Sheets, and Sea Level. Washington, D.C.: National Academy Press.

National Research Council, Carbon Dioxide Assessment Committee. 1983. Changing Climate. Washington, D.C.: National Academy Press.

National Research Council. 1982. Carbon Dioxide and Climate: A Second Assessment. Washington, D.C.: National Academy Press.

National Research Council. 1979. Carbon Dioxide and Climate: A Scientific Assessment. Washington, D.C.: National Academy Press.

Newman, W. S., and R. W. Fairbridge. 1986. The management of sea level rise. Nature 320:319–321.

North Carolina Office of Coastal Management. 1984. North Carolina Administrative Code, Chapter 7H, 1983. Raleigh: The Office of Coastal Management.

Nummedal, D., and J. W Sneden. 1987. Sediment exchange between the shoreface and the continental shelf—evidence from the modern Texas coast and the rock record. Pp. 2110–2125 *in* Coastal Sediments '87, Vol. 2. New York: American Society of Civil Engineers.

Nummedal, D., and S. Penland. 1981. Sediment dispersal in Nordermeyor Seegat, West Germany. Pp. 187–200 *in* Special Publication No. 5. Liege, Belgium: International Association of Sedimentologists.

O'Brien, M. P. 1969. Equilibrium flow areas of inlets on sandy coasts. J. Waterways and Harbors Div. ASCE 95(WW1).

Oertel, G. F. 1979. Barrier island development during Holocene recession, southeastern United States. Pp. 273–290 *in* Barrier Islands, S. P. Leatherman, ed. New York: Academic Press.

Oertel, G. F., J. E. Fowler, and J. Pope. 1985. History of Eroson and Erosion Control Efforts at Tybee Island, Georgia. CERC-85-1. Washington, D.C.: U.S. Army Corps of Engineers.

Office of Technology Assessment. 1984. Wetlands: Their Use and Regulation. Washington, D.C.: OTA.

Penland, S., J. R. Suter, and R. Boyd. 1985. Barrier island area along abandoned Mississippi River deltas. Mar. Geol. 63:197–233.

Perdikis, H. S. 1967. Hurricane flood protection in the United States. J. Waterways and Harbors Div. ASCE 93(WW1):1–24.

Pirazzoli, P. A. 1986. Late Quaternary sea-level changes and coastal evolution. Special Issue 1, J. Coastal Res., P. A. Pirazzoli and T. R. Suter, eds.

Pirazzoli, P. A. 1984. Secular trends of relative sea level (RSL) changes indicated by tide-gauge records. Mar del Plata (Argentina): Simposio Internacional sobre Cambios del Nivel del Mar. Abstracts 82–85.

Poland, J. F. 1970. Land subsidence and aquifer-system compaction, Santa Clara Valley, California. Pp. 285–294 *in* Land Subsidence: Proceedings of the Tokyo Symposium, September, 1969, Vol. I. Paris: UNESCO.

Pugh, J. C. 1954. A classification of the Nigerian coastline. J. West African Sci. Assoc. 1:3–12.

Pugh, D. T., and H. E. Faull. 1983. Operational Sea-Level Stations. Technical Series 23. Paris: UNESCO, Intergovernmental Oceanogr. Commission. 40 pp.

Ramanathan, V., R. J. Cicerone, H. B. Singh, and J. T. Kiehl. 1985. Trace gas trends and their potential role in climate change. J. Geophys. Res. 90(03):5547–5566.

Redfield, A. C. 1972. Development of a New England Salt Marsh. Ecol. Monogr. 42:201–237.

Revelle, R. 1983. Probable future changes in sea level resulting from increasing atmospheric carbon dioxide. *In* Changing Climate. Washington, D.C.: National Academy Press.

Robin, G. de Q. 1986. Changing sea level. *In* Greenhouse Effect, Climatic Change, and Ecosystems. New York: John Wiley & Sons. p. 323.

Rosen, P. S. 1978. A regional test of the Bruun rule on shoreline erosion. Mar. Geol. 26:7–16.

Schwartz, M. L. 1967. The Bruun theory of sea-level rise as a cause of shore erosion. J. Geol 73:528–534.

Schwartz, M. L. 1965. Laboratory study of sea level rise as a cause of shore erosion. J. Geol. 73:528–534.

Seymour, R. J., R. R. Strange, D. R. Cayan, and R. A. Rathan. 1984. Influence of El Niños in California's wave climate. Pp. 577–592 *in* Proceedings of Nineteenth Coastal Engineering Conference, September 3–7, 1984, Vol. I. New York: American Society of Civil Engineers.

Shalowitz, A. L. 1964. Shore and Sea Boundaries. U.S. Department of Commerce Pub. 10–1 (2):749.

Shepard, F. P. 1963. Submarine Geology. New York: Harper and Brothers.

Shepard, F. P., and H. R. Wanless. 1971. Our Changing Coastlines. New York: McGraw-Hill. 579 pp.

Steele, G. A. 1980. Stratigraphy and Depositional History of Bogue Banks, North Carolina. M.S. thesis. Duke University, Durham, North Carolina. 201 pp.

Stephens, J. C., and W. H. Speir. 1970. Subsidence of organic soils in the U.S.A. Pp. 523–534 *in* Proceedings of the Tokyo Symposium, September 1969, Vol. II. Paris: UNESCO.

Stevenson, J. C., L. G. Ward, and M. S. Kearney. 1986. Vertical accretion in marshes with varying rates of sea level rise. Pp. 241–260 *in* Estuarine Variability, D. Wolf, ed. New York: Academic Press.

Sullivan, W. 1986. Geologists warn of sea level rise. New York Times, April 11, 1986.

Sunamura, J. 1983. Processes of sea cliff and platform erosion. Pp. 233–265 *in* Handbook of Coastal Processes and Erosion. Boca Raton, Fla.: CRC Press.

Swift, D. J. P. 1975. Barrier island genesis: Evidence from the Central Atlantic shelf, eastern USA. Sed. Geol. 14:1–43.

Swift, D. J. P. 1968. Coastal erosion and transgressive stratigraphy. J. Geol. 76:445–456.

Swift, D. J. P., J. W. Kofoed, F. P. Saulsbury, and P. Sears. 1972. Holocene evolution of the shelf surface central and Southern Atlantic shelf of North America. Pp. 499–574 *in* Shelf Sediment Transport, D. J. P. Swift, D. B. Duane, and O. H. Pilkey, eds. Stroudsburg, Pa.: Dowder, Hutchinson and Ross.

Tagami, H., T. Kanno, K. Teranaka, and K. Kono. 1970. High tides countermeasures in land subsidence area. Pp. 622–628 *in* Land Subsidence: Proceedings of the Tokyo Symposium, September 1969, Vol. II. Paris: UNESCO.

Takeuchi, S., S. Kimoto, M. Wada, H. Shuna, and K. Markai. 1970. Geological and geohydrological properties of the land subsided areas. Pp. 232–241 *in* Land Subsidence: Proceedings of the Tokyo Symposium, September 1969. Paris: UNESCO.

Taney, N. E. 1961. Geomorphology of the South Shore of Long Island, New York. Beach Erosion Board, Technical Memorandum No. 128. Washington, D.C.: U.S. Army Corps of Engineers.

Tanner, W. F. 1960. Florida coast classification. Trans. Gulf Coast Assoc. Geol. Soc. 10:259–266.

Tanner, W. F., and F. Stapor. 1971. Tobasco Beach ridge plain: An eroding coast. Trans. Gulf Coast Assoc. Geol. Soc. 21:231–232.

Teichert, C. 1947. Contemporary eustatic rise in sea level? Geogr. J. 109:288–289.

Ters, M. 1986. Variations in Holocene sea level on the French Atlantic coast and their climatic significance. Pp. 87–103 *in* Climate: History, Periodicity, and Predictability, M. R. Ramkino, W. S. Newman, and J. E. Sanders, eds. New York: Van Nostrand Reinhold.

Thomas, R. H. 1985. Responses of the polar ice sheets to climatic warming. *In* Glaciers, Ice Sheets, and Sea Level. Washington, D.C.: National Academy Press.

Thorarinsson, S. 1940. Present glacier shrinkage and eustatic change in sea level. Geogr. Ann. 22:131-159.

Thornton, E. B., A. J. Sklavidts, W. L. Blanco, D. M. Burych, S. P. Tucker, and D. Puccini. 1985. Coastal erosion along southern Monterey Bay. Paper presented at the West Coast Regional Coastal Design Conference, sponsored by the U.S. Army Corps of Engineers and the ASCE, Oakland, California. November.

Titus, J. G. 1986. Greenhouse effect, sea level rise, and coastal zone management. CZMJ 14(3):147–171.

Titus, J. G. 1985. Sea level rise and the Maryland coast. *In* Potential Impacts of Sea Level Rise on the Beach at Ocean City, Maryland, J. G. Titus, ed. Washington, D.C.: U.S. Environmental Protection Agency.

Titus, J. G., C. Y. Kuo, M. J. Gibbs, T. B. LaRoche, M. K. Welts, and J. O. Waddell. 1987. Greenhouse effect, sea level rise, and coastal drainage systems. J. Water Res. Planning and Management. ASCE 113(2):216–227.

Tooley, M. T. 1978. Sea-Level Changes in Northwest England During the Flandrian Stage. Oxford: Oxford University Press. 232 pp.

Toyoshima, O. 1982. Variation of foreshore due to detached breakwaters. Pp. 1873–1982 *in* Coastal Engineering: Proceedings of the 18th Coastal Engineering Conference, Capetown, South Africa, Vol. III. New York: ASCE.

Ukena, K., T. Kanno, and K. Taranaka. 1970. Countermeasure against land subsidence with respect to displacement of surface flooding. Pp. 622–628 *in* Land Subsidence: Proceedings of the Tokyo Symposium, September 1969, Vol. II. Paris: UNESCO.

U.S. Army Corps of Engineers. 1984. Shore Protection Manual. Vicksburg, Miss.: Waterways Experiment Station.

U.S. Army Corps of Engineers. 1980. Beach Erosion Control and Storm Protection, Atlantic Coast of Maryland and Assateague Island, Virginia. Washington, D.C.: Department of the Army. 39 pp.

U.S. Army Corps of Engineers. 1981. Low Cost Shore Protection: Final Report on the Shoreline Erosion Control Demonstration Project (Section 54). Washington, D.C.: Department of the Army. 830 pp.

U.S. Army Corps of Engineers. 1971. National Shoreline Study. Washington, D.C.: Department of the Army. 59 pp.

Van De Plassche, O. 1986. Sea-Level Research: Manual for Collection and Evaluation of Data. Norwich: Geobooks. 600 pp.

van Veen, J. 1962. Dredge, Drain, Reclaim, The Art of a Nation, 5th ed. Dordrecht, Netherlands: Martinus Nijhoff. 200 pp.

Waddell, J. O., and R. A. Blaylock. 1987. Impact of Sea Level Rise on Gap Creek Watershed in the Fort Walton Beach, Florida, Area. Washington, D.C.: U.S. Environmental Protection Agency.

Wagret, P. 1968. Polderlands. London: Methuen & Co., Ltd. 288 pp.

Walton, T. L., and W. D. Adams. 1976. Capacity of inlet outer bars to store sand. Pp. 1919–1937 *in* Proceedings of the Fifteenth International Conference on Coastal Engineering. New York: American Society of Civil Engineers.

Weggel, R. 1979. A Method for Establishing Long-Term Erosion Rates from a Long-Term Rise in Water Level. CETA 79-2. Vicksburg, Miss.: U.S. Army Coastal Engineering Research Center. 16 pp.

Williams, P. B. 1985. An overview of the impact of accelerated sea level rise on San Francisco Bay. San Francisco: Phillip Williams and Associates. 25 pp.

Wright, H. E., ed. 1983. Late-Quaternary Environments of the United States: Vol. 2, The Holocene. Minneapolis: University of Minnesota Press. 277 pp.

Wyrtki, 1985. Sea-level data by satellite. EOS (August 6):578.

Zaremba, R. E., and S. P. Leatherman. 1986. Vegetative physiographic analysis of a U.S. north barrier system. Environ. Geol. Water Sci. 8:193–207.

Appendix A
Summary of Committee Expertise

ROBERT G. DEAN, chairman, is a coastal engineer specializing in tidal entrances, sand transport, and coastal nearshore processes. He has been department chairman, Department of Coastal and Oceanographic Engineering, University of Florida; professor of civil engineering at the Massachusetts Institute of Technology (MIT) and the University of Delaware; and senior research engineer, California Research Corporation. He is now professor of coastal and oceanographic engineering at the University of Florida and director of the Division of Beaches and Shores for the state of Florida. He holds B.S. and M.S. degrees from the University of California and Texas A&M University, and a Sc.D. in hydrodynamics from MIT. Dr. Dean served as chairman of the Marine Board Committee on Wave-Measurement Technologies; he is a member of the National Academy of Engineering and a former member of the Marine Board.

ROBERT A. DALRYMPLE is professor of civil engineering at the University of Delaware in the College of Engineering and College of Marine Studies, where he has been for 13 years. He holds a B.S. degree in engineering sciences from Dartmouth College, M.S. degree in ocean engineering from the University of Hawaii, and Ph.D. degree in civil and coastal engineering from the University of Florida. His research interests are in littoral processes and inlets, wave mechanics, and numerical modeling of nearshore processes. His consulting engineering has been in shoreline setback limits for development, beach nourishment, and wave-propagation modeling.

RHODES W. FAIRBRIDGE is professor emeritus of geology at Columbia University and adjunct professor at the State University of New York at Stony Brook. He has conducted and published research for many years in eustatic sea level change. His research interests also include the role of gravitational processes in tectonic change and sedimentation, paleoclimatology, and world geotectonics. Dr. Fairbridge has participated in several studies of the National Research Council. He has served as president of the Shorelines Commission and of the Neotectonic Commission of the International Union for Quarternary Research. His field experience and expeditions have taken him to most countries having coastlines. Editorial work includes the *Encyclopedia of Oceanography* and eight others, some 90 volumes of *Benchmark* collections, and assistance in founding the *Journal of Coastal Research.* He has been awarded the Alexander von Humbolt Prize and an honorary degree by the University of Gothenburg, Sweden.

STEPHEN P. LEATHERMAN is professor of geomorphology in the Department of Geography and director of the Laboratory for Coastal Research, University of Maryland. His principal research interests are in quantitative coastal geomorphology, coastal geology and hydraulics, and coastal resource management. His B.S. degree in geoscience is from North Carolina State University, and his Ph.D. in environmental sciences is from the University of Virginia. He has served as a consulting coastal geomorphologist on several national committees assessing the status and prospects for barrier beaches, and has published research in geomorphic responses of landforms to rising relative sea levels, migration of barrier islands, and mainland influences on coastal transgressions.

DAG NUMMEDAL is professor of geology at Louisiana State University. His research has concentrated on shallow marine sedimentation, particularly tidal inlet stability and tidal delta sedimentation, barrier island evolution, and shoreline change. His current research is focusing on sedimentation in modern and ancient continental shelves. He has served as consulting geologist to the National Aeronautics and Space Administration and is a current member of the Coastal Engineering Board of the U.S. Army

Corps of Engineers. Nummedal holds B.A. and M.A. degrees from the University of Oslo, Norway. His Ph.D. degree is from the University of Illinois.

MORROUGH P. O'BRIEN has been engaged throughout his long career in research, teaching, and practice in fluid mechanics and coastal engineering. He retired in 1959 as dean of engineering, University of California at Berkeley. In 1929, he inaugurated the first program of research on coastal engineering in the United States, under the sponsorship of the U.S. Army Corps of Engineers. He served as a member of the Beach Erosion Board (now the Coastal Engineering Research Board) for more than 40 years. He has been a member of the Army Scientific Advisory Panel, the Defense Science Board, and the National Science Board. Since 1949, he has served as a consultant to the management of Aerospace and Defense Groups, General Electric Company. He continues his work in coastal engineering as an adjunct professor at the University of Florida. His B.S. degree in civil engineering was granted by the Massachusetts Institute of Technology in 1925; he has received honorary degrees of D.Sc. (Northwestern), D. Eng. (Purdue), and LL.D. (University of California, Berkeley). He is a member of the National Academy of Engineering.

ORRIN H. PILKEY has been professor of marine geology at Duke University for almost 20 years. His research publications are concentrated in marine sedimentation, rising sea levels and shoreline sediments (particularly barrier islands), and the sedimentology of the continental shelves. Dr. Pilkey's B.S., M.S., and Ph.D. degrees in geology are from Washington State University, Montana State University, and Florida State University, respectively. He has served on National Research Council committees, as editor-in-chief of the *Journal of Sedimentary Petrology* (1978–1983), and as president of the Society of Economic Paleantologists and Mineralogists (1985–1986).

WILTON STURGES III is professor of oceanography at Florida State University. He has been a member of the Ocean Sciences Board and several committees of the National Research Council. His research interests are in sea-surface topography, particularly sea-surface slopes, thermal expansion, and ocean mixing and circulation. His B.S. degree in physics is from Auburn University, and his M.S. and Ph.D. degrees in oceanography are from the Johns Hopkins University.

ROBERT L. WIEGEL is professor of civil engineering at the University of California at Berkeley, where he has been employed since June 1946. His professional activities concentrate on coastal and ocean engineering, with particular attention to the response of ocean and coastal structures to environmental forces and to the development of environmental design criteria. He was founding president of the United Nations Engineering Committee on Oceanic Resources (advisory to UNESCO), and served for 6 years as a member of the Marine Board, as well as a member and chairman of several Marine Board studies. Professor Wiegel (whose degrees in mechanical engineering are from the University of California, Berkeley) is a member of the National Academy of Engineering.

Appendix B

ACKNOWLEDGMENTS

The committee gratefully acknowledges the generous contributions of time and information provided by the liaison representatives and their agencies, and the many individuals who participated in the data-gathering processes inherent to the project. Special thanks are extended to all who communicated with the project by telephone and mail, as well as to the following individuals who participated in committee meetings and workshops. The cooperation and interest in the committee's work of all respondents was of material assistance.

Sherburne Abbott, National Research Council
David Aubrey, Woods Hole Oceanographic Institution
David Barilovich, U.S. Army Corps of Engineers
T. Barnett, Scripps Institution of Oceanography
L. Borgman, University of Wyoming
Kirk Bryan, Princeton University
J. Collins, American Society of Photogrammetry
K. O. Emery, Woods Hole Oceanographic Institution
D. Foster, Australia

Cyril Galvin, Springfield, Virginia
Vivien Gornitz, Goddard Institute for Space Studies
James E. Hansen, Goddard Institute for Space Studies
Eugene H. Harlow, Soros Associates
Steacy D. Hicks, National Oceanic and Atmospheric Administration
H. Horikawa, Japan
John G. Housley, U.S. Army Corps of Engineers
C. H. J. Hull, Delaware River Basin Commission
Douglas James, Utah State University
Michael Kearney, University of Maryland
George Kukla, Lamont-Doherty Geological Observatory
Nicholas Kraus, U.S. Army Corps of Engineers
Ray B. Krone, University of California, Davis
Ashish J. Mehta, University of Florida
Orville Magoon, Guenoc Winery
Walter Newman, Queens College
Joan Pope, U.S. Army Corps of Engineers
Frederick Sanders, Massachusetts Institute of Technology
Wolfgang Scherer, National Oceanic and Atmospheric Administration
Hsieh Wen Shen, University of California, Berkeley
Robert Sorensen, Lehigh University
F. W. Stapor, Tennessee Technical University
C. Thurlow, National Oceanic and Atmospheric Administration
B. Thom, Australia
Thomas M. Usselman, National Research Council
Lawrence Ward, Horn Paint Environmental Laboratory
R. Weggel, Drexel University
John Wells, University of North Carolina
Peter K. Weyl, State University of New York, Stony Brook
Robert W. Whalin, U.S. Army Corps of Engineers

Index

F

G

H

I

J

L